D0081691

Cambridge Tracts in Mathematics
and Mathematical Physics

GENERAL EDITORS
H. BASS, J. F. C. KINGMAN, F. SMITHIES,
J. A. TODD AND C. T. C. WALL.

No. 43

AN INTRODUCTION TO HOMOTOPY THEORY

AN INTRODUCTION TO HOMOTOPY THEORY

BY

P. J. HILTON

M.A., D.PHIL., PH.D.

Professor of Mathematics
Cornell University

CAMBRIDGE
AT THE UNIVERSITY PRESS
1953

WILLIAM MADISON RANDALL LIBRARY UNC AT WILMINGTON

Published by the Syndics of the Cambridge University Press
Bentley House, 200 Euston Road, London, NW1 2DB
American Branch: 32 East 57th Street, New York, N.Y.10022

ISBN: 0 521 05265 3

First published 1953
Reprinted 1961 1964 1966 1972

First printed in Great Britain at the University Press, Cambridge
Reprinted in Great Britain by John Dickens & Co Ltd, Northampton

QA611
.H65

CONTENTS

139295

Chapter VI. THE HOPF INVARIANT AND SUSPENSION THEOREMS

Chapter VII. WHITEHEAD CELL-COMPLEXES

Chapter VIII. HOMOTOPY GROUPS OF COMPLEXES

PREFACE

Since the introduction of homotopy groups by Hurewicz in 1935, homotopy theory has been occupying an increasingly prominent place in the field of algebraic topology. Important new advances are continually being made in the subject by various workers; and the recent developments emanating from the French school of topologists underline the desirability of having available a basic introduction to homotopy theory suitable for those who wish to undertake research in the subject and for those who wish to be in a position to understand the modern techniques and results.

At the moment, no text-book of homotopy theory exists at any level, with the result that the newcomer to this branch of mathematics is obliged to plunge straight into the study of original papers, often of very considerable complexity. This monograph is designed to fill the gap. It does not claim to be a comprehensive treatment of its subject (the recent work of the French school, for example, is not included, except for a brief introduction to it in the last section of Chapter V); but it is hoped that a reader familiar with Lefschetz's *Introduction to Topology* will obtain an understanding of the fundamental ideas of homotopy theory from the first six chapters of this book.

The final two chapters are somewhat different in kind from the first six, being an account of the homotopy theory of complexes. J. H. C. Whitehead's generalization of the simplicial complex has been very fruitful, and it seems proper that a book of this kind should contain a brief treatment of it. Chapter VII is devoted to such a treatment, while Chapter VIII offers the reader a particular application of the results of Chapter VII and earlier chapters.

The tract ends with a bibliography and index-glossary. This latter is designed to provide the reader with brief definitions of the fundamental terms used in homotopy theory without obliging him to consult the relevant parts of the text.

The author wishes to express his gratitude to Prof. M. H. A. Newman and Mr D. B. Scott for reading the manuscript and suggesting many improvements. He is also greatly indebted to Mr S. Wylie for making valuable criticisms and suggestions at the proof-reading stage.

Finally, the author wishes to acknowledge the kind co-operation of the Cambridge University Press and their skilful preparation of this volume.

P. J. H.

Victoria University of Manchester
June 1952

HOMOTOPY THEORY

CHAPTER I

INTRODUCTION

Let X, Y be two arcwise-connected Hausdorff spaces and consider maps (i.e. continuous transformations) of X into Y. We say that $f_0, f_1: X \rightarrow Y$ are *homotopic* if there exists a map $F: X \times I \rightarrow Y$, where I is the unit interval $0 \leqslant t \leqslant 1$, such that $F(x, 0) = f_0(x)$, $F(x, 1) = f_1(x)$ for all $x \in X$. This may also be expressed by saying that there exists a continuous family of maps $f_t: X \rightarrow Y$, $0 \leqslant t \leqslant 1$, but it must be remembered that the continuity is with respect to $x \in X$ and $t \in I$ simultaneously. If f_0 is homotopic to f_1, we write $f_0 \sim f_1: X \rightarrow Y$, or, where no confusion may arise, $f_0 \sim f_1$.

Suppose that, under both f_0 and f_1, a certain subset $X_0 \subset X$ is mapped into $Y_0 \subset Y$. Then we write $f_0, f_1: X, X_0 \rightarrow Y, Y_0$. We may restrict the homotopy by insisting that, throughout the deformation from f_0 to f_1, the image of X_0 is to remain in Y_0. If f_0 is homotopic to f_1 under such a deformation, we write

$$f_0 \sim f_1: X, X_0 \rightarrow Y, Y_0.$$

We may consider homotopies under which the image of a certain subset of X remains pointwise fixed. If $f_0, f_1: X \rightarrow Y$ and† if $f_0 | X_0 = f_1 | X_0$ for some $X_0 \subset X$, then we write $f_0 \sim f_1: X \rightarrow Y$, rel X_0, or, briefly, $f_0 \sim f_1$, rel X_0, if there exists a map $F: X \times I \rightarrow Y$ such that $F(x, 0) = f_0(x)$, $F(x, 1) = f_1(x)$, $x \in X$, $F(x_0, t) = f_0(x_0)$, $x_0 \in X_0$, $t \in I$.

We now prove a fundamental lemma in the construction of continuous functions.

† By $f | X_0$, where f is a map $X \rightarrow Y$ and $X_0 \subset X$, we mean the map $f': X_0 \rightarrow Y$, given by $f'(x_0) = f(x_0)$, $x_0 \in X_0$. We call $f | X_0$ the restriction of f to X_0.

LEMMA 1·1. *Let the space X be the union of two closed subspaces A and B. Let $f: A \to Y$, $g: B \to Y$ be maps such that*

$$f \mid A \cap B = g \mid A \cap B.$$

Then the transformation $h: X \to Y$, given by $h \mid A = f$, $h \mid B = g$, is continuous.

For let F be any closed set in Y. Now clearly

$$h^{-1}(F) = f^{-1}(F) \cup g^{-1}(F).$$

Since f is continuous, $f^{-1}(F)$ is closed in A; but A is closed in X, so that $f^{-1}(F)$ is closed in X. Similarly, $g^{-1}(F)$ is closed in X, so that $h^{-1}(F)$ is closed in X, whence h is continuous.

THEOREM 1·2. *The relations*

$$f_0 \sim f_1 : X \to Y, \quad f_0 \sim f_1 : X, X_0 \to Y, Y_0, \quad f_0 \sim f_1 : X \to Y, \operatorname{rel} X_0$$

are equivalence relations.

That the relations are symmetrical and reflexive is trivial. We prove transitivity. Let $f' \sim f'' : X \to Y$, $f'' \sim f''' : X \to Y$. Then there exist $F_1 : X \times I \to Y$, $F_2 : X \times I \to Y$, such that

$$F_1(x, 0) = f'(x), \quad F_1(x, 1) = F_2(x, 0) = f''(x), \quad F_2(x, 1) = f'''(x).$$

Define $F : X \times I \to Y$ by

$$\begin{aligned} F(x, t) &= F_1(x, 2t), & 0 \leqslant t \leqslant \tfrac{1}{2} \\ &= F_2(x, 2t - 1), & \tfrac{1}{2} \leqslant t \leqslant 1. \end{aligned}$$

By lemma 1·1 F is continuous and it clearly furnishes a homotopy between f' and f'''. Moreover, if F_1 and F_2 were restricted. F would be similarly restricted.

Theorem 1·2 enables us to divide maps of X into Y, or restricted maps of X into Y, into *homotopy classes*. The central problem of homotopy theory is to express the properties of the collection of homotopy classes of maps of X into Y in terms of topological invariants† of X and of Y.

Let $f: X \to Y$ be a map such that $f(X)$ is a single point of Y. We then call f a *constant* map. Since Y is arcwise-connected, any

† It is clear that if X', Y' are homeomorphic to X, Y respectively, then the homotopy classes of maps $X \to Y$ may be set in (1-1) correspondence with the homotopy classes of maps $X' \to Y'$.

two constant maps are homotopic, so that we may refer to the class of the constant map.

THEOREM 1·3. *Let E^n be an n-element (the homeomorph of the interior and boundary of a sphere in n-dimensional Euclidean space). Then any map of E^n into Y is homotopic to the constant map.*

Let E^n be taken as the full unit sphere, centre the origin, in n-space. Then any point of E^n may be represented as λx, where $x \in S^{n-1}$, the surface of E^n, and $0 \leqslant \lambda \leqslant 1$. Define $F: E^n \times I \to Y$ by $F(\lambda x, t) = f((1-t)\lambda x)$, where $f: E^n \to Y$ is any map. Then $F(\lambda x, 0) = f(\lambda x)$ and $F(\lambda x, 1) = f(0)$, where 0 is the origin. Of course, the essential fact which we use here is the fact that E^n is contractible over itself to a point.

To establish non-triviality, we mention at this stage that the identity map of the circle on itself is not homotopic to the constant map. In fact there exists an enumerable infinity of classes of maps of the circle into itself.†

Now let $f: X \to Y$, $g: Y \to X$ be such that $gf: X \to X$ and $fg: Y \to Y$ are both homotopic to the appropriate identity maps. Then we say that X and Y are of the *same homotopy type*. We call f and g *homotopy equivalences* and $g(f)$ a *homotopy inverse* of $f(g)$. We write 1 for the identity map of any space on itself, so that $gf \sim 1$, $fg \sim 1$. If X and Y are of the same homotopy type, we write $X \sim Y$.

THEOREM 1·4. *The relation $X \sim Y$ is an equivalence relation.*

That the relation is symmetrical and reflexive is trivial. To prove transitivity (and in several other contexts in homotopy theory), we require the following lemma:

LEMMA 1·5. *Let $f_0 \sim f_1: X \to Y$, $g: Y \to W$, $h: Z \to X$. Then $gf_0 \sim gf_1: X \to W$, $f_0 h \sim f_1 h: Z \to Y$.*

For let $F: X \times I \to Y$ be a homotopy between f_0 and f_1. Then $gF: X \times I \to W$ is a homotopy between gf_0 and gf_1; and

$$F^*: Z \times I \to Y, \quad \text{given by} \quad F^*(z, t) = F(h(z), t), \quad z \in Z,\ t \in I,$$

is a homotopy between $f_0 h$ and $f_1 h$.

† For an elementary proof, see M. H. A. Newman, *Topology of Plane Sets*, 2nd edition, C.U.P. (1951), Chapter VII. It is shown there that, if 0 is the origin of Euclidean 2-space R^2, then two loops in $R^2 - 0$ are homotopic in $R^2 - 0$ if and only if the *order* of 0 with respect to each is the same, and the *order* ('the winding number round 0') can take any integer value.

Reverting to the theorem, let $f: X \to Y$, $g: Y \to X$ be such that $gf \sim 1: X \to X$, $fg \sim 1: Y \to Y$, and let $u: Y \to Z$, $v: Z \to Y$ be such that $vu \sim 1: Y \to Y$, $uv \sim 1: Z \to Z$. Consider the maps $uf: X \to Z$, $gv: Z \to X$. By the lemma, since $vu \sim 1: Y \to Y$, $gvu \sim g: Y \to X$, and $gvuf \sim gf: X \to X$. But $gf \sim 1: X \to X$, so that, by theorem 1·2, $gvuf \sim 1: X \to X$. Similarly, $ufgv \sim uv \sim 1: Z \to Z$. Thus uf is a homotopy equivalence of X to Z and gv is a homotopy inverse of uf.

It is clear that homeomorphic spaces are of the same homotopy type, so that invariants of homotopy type are, *a fortiori*, topological invariants. However, the example of $X = E^n$, $Y = y_0$, a single point, shows that two spaces may be of the same homotopy type without being homeomorphic. Among the invariants of homotopy type are the singular homology groups and the *homotopy groups*; these latter form the subject-matter of the next chapter.

CHAPTER II

THE HOMOTOPY GROUPS

1. Definition of the absolute homotopy groups. Let I^n be the Euclidean n-cube, consisting of points $(t_1, \ldots, t_n)$, with $0 \leqslant t_i \leqslant 1$, $i = 1, \ldots, n$. Actually it is convenient to regard I^n as embedded in Hilbert space, so that, strictly speaking, the co-ordinates of points of I^n are $(t_1, \ldots, t_n, 0, \ldots)$. In this way I^m is identified with a particular face of I^n if $m < n$. However, where no confusion can arise, we will omit the string of zeros. The boundary of I^n, written $\dot{I}^n$, consists of those points of I^n for which $t_i = 0$ or 1 for at least one value of i.

Now let Y be an arcwise-connected Hausdorff space and let $y_0 \in Y$. Let $f, g : I^n \to Y$ be such that $f(1, t_2, \ldots, t_n) = g(0, t_2, \ldots, t_n)$ and define $h : I^n \to Y$ by

$$\begin{aligned} h(t_1, t_2, \ldots, t_n) &= f(2t_1, t_2, \ldots, t_n), & 0 \leqslant t_1 \leqslant \tfrac{1}{2} \\ &= g(2t_1 - 1, t_2, \ldots, t_n), & \tfrac{1}{2} \leqslant t_1 \leqslant 1. \end{aligned}$$

Then h is continuous and we write $h = f + g$.

Suppose now that f, g both map $\dot{I}^n$ to y_0, and let $M_n(Y, y_0)$ stand for the totality of maps $I^n, \dot{I}^n \to Y, y_0$. Then $f + g$ is defined and belongs to $M_n(Y, y_0)$. Thus we have introduced an addition into the collection $M_n(Y, y_0)$. Let $\pi_n(Y, y_0)$ stand for the totality of homotopy classes of maps $I^n, \dot{I}^n \to Y, y_0$, and let $[f]$ be the class of f. Then we may show that if $f, g \in M_n(Y, y_0)$, then $[f+g]$ depends only on $[f]$ and $[g]$. For suppose $f \simeq f', g \simeq g' : I^n, \dot{I}^n \to Y, y_0$. Then there exist $F, G : I^n \times I$, $\dot{I}^n \times I \to Y, y_0$ with $F(x, 0) = f(x)$, $F(x, 1) = f'(x)$, $G(x, 0) = g(x)$, $G(x, 1) = g'(x)$, $x \in I^n$. H, $= F + G$, is defined and is a map $H : I^n \times I$, $\dot{I}^n \times I \to Y, y_0$. It is, moreover, clear that H is a homotopy between $f + g$ and $f' + g'$. Thus the addition introduced into $M_n(Y, y_0)$ induces an addition in $\pi_n(Y, y_0)$.

Of course, the addition defined in $\pi_n(Y, y_0)$ is not only describable in terms of the particular addition in $M_n(Y, y_0)$. It will be convenient for future applications to give the following more general description.

Let I_1^n be the 'subcube' of I^n consisting of points $(t_1, \dots, t_n)$ with $\lambda_i \leqslant t_i \leqslant \mu_i$, where $0 \leqslant \lambda_i < \mu_i \leqslant 1$, $i = 1, \dots, n$.

LEMMA 1·1. *Given* $f \in M_n(Y, y_0)$, *there exists* $f' \in M_n(Y, y_0)$ *such that* $f' \in [f]$ *and* $f'(I^n - I_1^n) = y_0$.

For let $f_t \in M_n(Y, y_0)$ be given by

$$f_t(t_1, \dots, t_n) = f\left(\frac{t_1 - \lambda_1 t}{1-(1-\mu_1+\lambda_1)t}, \dots, \frac{t_n - \lambda_n t}{1-(1-\mu_n+\lambda_n)t}\right)$$
$$\text{if } \lambda_i t \leqslant t_i \leqslant 1-(1-\mu_i)t, \text{ for all } i,$$
$$= y_0, \quad \text{otherwise.}$$

Then $f_0 = f$ and $f_1(t_1, \dots, t_n) = y_0$ unless $\lambda_i \leqslant t_i \leqslant \mu_i$, all i. Put $f' = f_1$. We say that the map f' results from *concentrating* f on I_1^n.

Now let $f, g \in M_n(Y, y_0)$ and let f', g' result from concentrating f, g on subcubes I_1^n, I_2^n respectively. Suppose further that the interiors of I_1^n and I_2^n are disjoint and that I_1^n lies 'to the left of' I_2^n, i.e. if I_1^n is as above and if I_2^n is given by $\lambda_i' \leqslant t_i \leqslant \mu_i'$, $i = 1, \dots, n$, then $\mu_1 \leqslant \lambda_1'$. Then a map $k \in M_n(Y, y_0)$ is given by

$$k \mid I_1^n = f', \quad k \mid I_2^n = g', \quad k(x) = y_0 \quad \text{if} \quad x \in I^n - (I_1^n \cup I_2^n).$$

LEMMA 1·2. $k \in [f+g]$.

We first prove the lemma if I_1^n, I_2^n are the half-cubes given by $0 \leqslant t_1 \leqslant \frac{1}{2}$, $\frac{1}{2} \leqslant t_1 \leqslant 1$, respectively. If we choose the particular concentrations given by the construction of lemma 1·1, we see that $k = f + g$. Now let f_1', g_1' be any concentrations of f, g on I_1^n, I_2^n respectively. Then $f_1' \in [f]$, $g_1' \in [g]$. Define $f_1 \in M_n(Y, y_0)$ by $f_1(t_1, t_2, \dots, t_n) = f_1'(\frac{1}{2}t_1, t_2, \dots, t_n)$ and define g_1 similarly. Then f_1', g_1' are the concentrations of f_1, g_1 given by lemma 1·1. Thus if k_1 is given by $k_1 \mid I_1^n = f_1'$, $k_1 \mid I_2^n = g_1'$, we have $k_1 = f_1 + g_1$, and $f_1 \in [f_1'] = [f]$, $g_1 \in [g_1'] = [g]$. Thus $k_1 \in [f+g]$. This proves the lemma in the special case.

Reverting to the general case, there is no loss of generality in supposing $\mu_1 = \lambda_1'$; for if $\mu_1 < \lambda_1'$ and f' concentrates f on I_1^n, then, *a fortiori*, f' concentrates f on the larger subcube obtained† by replacing μ_1 by λ_1'. We suppose then that $\mu_1 = \lambda_1'$. By the same

† Notice also that the definition of the map k is unaffected by this device.

token we may also suppose that $\lambda_1 = 0, \mu'_1 = 1$, and that $\lambda_i = \lambda'_i = 0$, $\mu_i = \mu'_i = 1$, $i \neq 1$. Now define $\sigma_t : I^1 \to I^1$ by

$$\begin{aligned}\sigma_t(t_1) &= t_1(1 + t(2\mu - 1)), & 0 \leqslant t_1 \leqslant \tfrac{1}{2}\\ &= 1 - (1 - t_1)(1 - t(2\mu - 1)), & \tfrac{1}{2} \leqslant t_1 \leqslant 1,\end{aligned}$$

where $\mu = \mu_1 = \lambda'_1$. Then σ_t is a deformation of I^1, rel $\dot{I}^1$, which deforms $\langle 0, \frac{1}{2}\rangle$ on to $\langle 0, \mu\rangle$ and $\langle \frac{1}{2}, 1\rangle$ on to $\langle \mu, 1\rangle$. Define

$$\bar{\sigma}_t : I^n, \dot{I}^n \to I^n, \dot{I}^n \quad \text{by} \quad \bar{\sigma}_t(t_1, t_2, \ldots, t_n) = (\sigma_t(t_1), t_2, \ldots, t_n).$$

Since $\bar{\sigma}_0 = 1$, $k\bar{\sigma}_1 \in [k]$, $f'\bar{\sigma}_1 \in [f'] = [f]$ and $g'\bar{\sigma}_1 \in [g'] = [g]$. Now $f'\bar{\sigma}_1$ concentrates f on the half-cube $0 \leqslant t_1 \leqslant \frac{1}{2}$, and $g'\bar{\sigma}_1$ concentrates g on the half-cube $\frac{1}{2} \leqslant t_1 \leqslant 1$. Moreover, $k\bar{\sigma}_1$ is given by $k\bar{\sigma}_1 \mid (\text{half-cube } 0 \leqslant t_1 \leqslant \frac{1}{2}) = f'\bar{\sigma}_1$, $k\bar{\sigma}_1 \mid (\text{half-cube } \frac{1}{2} \leqslant t_1 \leqslant 1) = g'\bar{\sigma}_1$. Thus, by the special case, $k\bar{\sigma}_1 \in [f+g]$; but $k\bar{\sigma}_1 \in [k]$, so that $k \in [f+g]$.

LEMMA 1·3. *$[f+g] = [g+f]$ if $n > 1$.*

Since I^2 is homeomorphic to a circular disk, we may define a rotation of I^2 through 180°. This rotation

$$\rho' : I^2, \dot{I}^2 \to I^2, \dot{I}^2$$

is homotopic to the identity and interchanges $t_1 \leqslant \frac{1}{2}$ and $t_1 \geqslant \frac{1}{2}$. Define $\rho : I^n, \dot{I}^n \to I^n, \dot{I}^n$ by

$$\rho(t_1, t_2, t_3, \ldots, t_n) = (\rho'(t_1, t_2), t_3, \ldots, t_n), \quad n > 1.$$

Then $\rho \sim 1$. Let f be concentrated on the half-cube $0 \leqslant t_1 \leqslant \frac{1}{2}$, g on the half-cube $\frac{1}{2} \leqslant t_1 \leqslant 1$ (we do not need to write f', g'). Then $f\rho$ is concentrated on the half-cube $\frac{1}{2} \leqslant t_1 \leqslant 1$, $g\rho$ on the half-cube $0 \leqslant t_1 \leqslant \frac{1}{2}$. Let us agree to call the half-cube $0 \leqslant t_1 \leqslant \frac{1}{2}$ I^n_L and the half-cube $\frac{1}{2} \leqslant t_1 \leqslant 1$ I^n_R. Then if k is given by $k \mid I^n_L = f$, $k \mid I^n_R = g$, we have $k \in [f+g]$ and $k\rho \mid I^n_L = g\rho$, $k\rho \mid I^n_R = f\rho$. Thus $k\rho \in [g\rho + f\rho]$. Since $\rho \sim 1$, it follows that $k\rho \in [f+g]$ and $g\rho + f\rho \in [g+f]$.

THEOREM 1·4. *Under the operation of addition, the collection of classes, $\pi_n(Y, y_0)$, is a group. It is Abelian if $n > 1$.*

(i) *Existence of right zero.* Let g be the constant map, $g(I^n) = y_0$, let $\alpha \in \pi_n(Y, y_0)$ and let $f \in \alpha$ be concentrated on I^n_L. Then the map k, given by $k \mid I^n_L = f$, $k \mid I^n_R = g$, is just f, so that $\alpha + [g] = \alpha$. We write $[g] = 0$.

(ii) *Existence of right inverse.* Let $\alpha \in \pi_n(Y, y_0)$ be represented by $f \in M_n(Y, y_0)$. Define $(-\alpha)$ as the class of the map $\bar{f}$, given by $\bar{f}(t_1, t_2, \ldots, t_n) = f(1-t_1, t_2, \ldots, t_n)$. It is clear that $(-\alpha)$ depends only on α. Then $\alpha + (-\alpha)$ is represented by $k \in M_n(Y, y_0)$, given by

$$\begin{aligned} k(t_1, t_2, \ldots, t_n) &= f(2t_1, t_2, \ldots, t_n), & 0 \leqslant t_1 \leqslant \tfrac{1}{2} \\ &= f(2-2t_1, t_2, \ldots, t_n), & \tfrac{1}{2} \leqslant t_1 \leqslant 1. \end{aligned}$$

Now define $g_t \in M_n(Y, y_0)$ by

$$\begin{aligned} g_t(t_1, t_2, \ldots, t_n) &= f(2(1-t)\, t_1, t_2, \ldots, t_n) & 0 \leqslant t_1 \leqslant \tfrac{1}{2} \\ &= f(2(1-t)\,(1-t_1), t_2, \ldots, t_n) & \tfrac{1}{2} \leqslant t_1 \leqslant 1. \end{aligned}$$

Then $g_0 = k$ and $g_1(I^n) = y_0$, so that $\alpha + (-\alpha) = 0$.

(iii) *Associativity.* Let $\alpha, \beta, \gamma \in \pi_n(Y, y_0)$, let f, g, h represent α, β, γ respectively, and let them be concentrated on the subcubes $0 \leqslant t_1 \leqslant \frac{1}{3}$, $\frac{1}{3} \leqslant t_1 \leqslant \frac{2}{3}$, $\frac{2}{3} \leqslant t_1 \leqslant 1$ respectively. It now follows from lemma 1·2 that the map which agrees with f on the first subcube, with g on the second, and with h on the third, represents both $(\alpha + \beta) + \gamma$ and $\alpha + (\beta + \gamma)$.

(iv) *Commutativity if* $n > 1$. This is just lemma 1·3.

The group $\pi_n(Y, y_0)$ is called the *nth* (*absolute*) *homotopy group of* Y mod y_0. If $n = 1$, it is just the fundamental group of Y based on loops with end-point y_0.

2. Alternative descriptions of the homotopy groups. Let us choose a fixed homeomorphism of I^n on to the standard n-element E^n, given by $x_1^2 + \ldots + x_n^2 \leqslant 1$. The boundary of E^n is the unit $(n-1)$-sphere, S^{n-1}, in Euclidean n-space, given by $x_1^2 + \ldots + x_n^2 = 1$. Using the given homeomorphism of I^n, $\dot{I}^n$ on to E^n, S^{n-1}, we may set up a (1-1) correspondence between elements of $\pi_n(Y, y_0)$ and homotopy classes of maps $E^n, S^{n-1} \to Y, y_0$.

Similarly, if E_+^n, given by $x_{n+1} \geqslant 0$, and E_-^n, given by $x_{n+1} \leqslant 0$, are the northern and southern hemispheres of S^n, the unit n-sphere in $(n+1)$-space, then each is an n-element bounded by S^{n-1}, and we may set up a (1-1) correspondence between elements of $\pi_n(Y, y_0)$ and homotopy classes of maps $E_+^n, S^{n-1} \to Y, y_0$ (or classes of maps $E_-^n, S^{n-1} \to Y, y_0$). Now let $p_0 \in S^n$ be the point $(1, 0, \ldots, 0)$.

THEOREM 2·1. *A (1-1) correspondence may be established between elements of* $\pi_n(Y, y_0)$ *and homotopy classes of maps*

$$S^n, p_0 \to Y, y_0.$$

By the remarks above, it is sufficient to set up a (1-1) correspondence between classes of maps $E^n_+, S^{n-1} \to Y, y_0$ and classes of maps $S^n, p_0 \to Y, y_0$. In fact, such a correspondence is set up by any map $\phi_n : E^n_+, S^{n-1} \to S^n, p_0$ which maps $E^n_+ - S^{n-1}$ homeomorphically on to $S^n - p_0$. For the correspondence $f \to f\phi_n$, $f : S^n, p_0 \to Y, y_0$, induces a correspondence of homotopy classes. If $g : E^n_+, S^{n-1} \to Y, y_0$ is a given map, then $g\phi_n^{-1} : S^n, p_0 \to Y, y_0$ is single-valued. It is therefore continuous. For, if F is closed in Y, $g^{-1}(F)$ is closed in E^n_+, so that $(g\phi_n^{-1})^{-1}(F), = \phi_n(g^{-1}(F))$, is closed† in S^n. Since $g = (g\phi_n^{-1})\,\phi_n$, it follows that $f \to f\phi_n$ induces a mapping of classes of maps $S^n, p_0 \to Y, y_0$ on to classes of maps

$$E^n_+, S^{n-1} \to Y, y_0.$$

To show that the mapping is (1-1), let $f, f' : S^n, p_0 \to Y, y_0$ and let $G : E^n_+ \times I, S^{n-1} \times I \to Y, y_0$ be such that

$$G(x, 0) = f\phi_n(x), \quad G(x, 1) = f'\phi_n(x), \quad x \in E^n_+.$$

Define $F : S^n \times I, p_0 \times I \to Y, y_0$ by $F(x, t) = G(\phi_n^{-1}(x), t)$, $x \in S^n$, $t \in I$. Then F is single-valued and $F(x, 0) = f(x)$, $F(x, 1) = f'(x)$. The continuity of F follows as above, with g, ϕ_n, S^n replaced by G, Φ_n, $S^n \times I$, where $\Phi_n : E^n_+ \times I \to S^n \times I$ is given by

$$\Phi_n(x, t) = (\phi_n(x), t), \quad x \in E^n_+, t \in I.$$

A suitable map for ϕ_n is the map rh, where r is the rotation $r(x_1, \ldots, x_{n+1}) = (-x_{n+1}, x_2, \ldots, x_n, x_1)$ and $h : E^n_+, S^{n-1} \to S^n, q_0$ is given by $h(x_1, \ldots, x_{n+1}) = (\mu x_1, \ldots, \mu x_n, 2x_{n+1} - 1)$, $\mu \geqslant 0$ being so chosen that $h(x) \in S^n$, $x \in E^n_+$. Here q_0 is the south pole $(0, 0, \ldots, -1)$ and E^n_+ is given‡ by $x_1^2 + \ldots + x_{n+1}^2 = 1$, $x_{n+1} \geqslant 0$. We will see in the next chapter that the (1-1) correspondence depends only on the orientation-class of the homeomorphism ϕ_n. In fact any two maps $\phi_n, \phi'_n : E^n_+, S^{n-1} \to S^n, p_0$ in the same orientation-class are homotopic.

† E^n_+ is, of course, compact.

‡ The author is indebted to Prof. M. H. A. Newman for this simple example of a suitable map. The map h may be extended to the whole of S^n by defining $h(E^n_-) = q_0$. It is then not difficult to see that h is homotopic (as a map $S^n, E^n_-, q_0 \to S^n, E^n_-, q_0$) to the identity. (It is the operation of 'closing a purse').

Any map $g: E^n_+, S^{n-1} \to Y, y_0$ may be extended to a map $g': S^n, p_0 \to Y, y_0$ by putting $g'(E^n_-) = y_0$. It may be shown that the correspondence $g \to g'$ induces a (1-1) correspondence of the appropriate homotopy classes of maps.

Let $\alpha \in \pi_n(Y, y_0)$ be represented by a map $f \in M_n(Y, y_0)$ which is concentrated on some subcube, I^n_1, of I^n. Then a map

$$g: I^n, \dot{I}^n \to S^n, p_0$$

which maps $I^n - \dot{I}^n$ homeomorphically on to $S^n - p_0$ induces a (1-1) correspondence of elements of $\pi_n(Y, y_0)$ with classes of maps $S^n, p_0 \to Y, y_0$. The map $fg^{-1}: S^n, p_0 \to Y, y_0$ will be concentrated on some closed n-cell, E^n_1, of S^n, where $g(I^n_1) = E^n_1$, provided only that I^n_1 is a proper subcube of I^n. Moreover, if $f_1, f_2 \in M_n(Y, y_0)$ represent $\alpha_1, \alpha_2 \in \pi_n(Y, y_0)$ and f_1, f_2 are concentrated on subcubes I^n_1, I^n_2 whose interiors are disjoint, then $f_1 g^{-1}$ and $f_2 g^{-1}$ will be concentrated on n-cells E^n_1, E^n_2 of S^n whose interiors are disjoint. If $k \in M_n(Y, y_0)$ is given by $k \mid I^n_1 = f_1$, $k \mid I^n_2 = f_2$, $k(x) = y_0$ otherwise, $x \in I^n$, then $k \in \alpha_1 + \alpha_2$, and $kg^{-1}: S^n, p_0 \to Y, y_0$ is given by $kg^{-1} \mid E^n_1 = f_1 g^{-1}$, $kg^{-1} \mid E^n_2 = f_2 g^{-1}$, $kg^{-1}(x) = y_0$ otherwise, $x \in S^n$. This shows how the addition in $\pi_n(Y, y_0)$ induces an addition in the classes of maps $S^n, p_0 \to Y, y_0$. Given two such classes β_1, β_2, we represent them by maps f_1, f_2, concentrated on n-cells E^n_1, E^n_2 whose interiors are disjoint and then define $k: S^n, p_0 \to Y, y_0$ by $k \mid E^n_1 = f_1$, $k \mid E^n_2 = f_2$, $k(x) = y_0$ otherwise, $x \in S^n$. Then k represents $\beta_1 + \beta_2$. If $n = 1$, we further insist that E^1_1 precedes E^1_2 in the given orientation of S^1, starting from the base-point.

The operation of addition then turns the collection of classes into a group isomorphic with $\pi_n(Y, y_0)$ if $n > 1$, and isomorphic or anti-isomorphic with $\pi_1(Y, y_0)$ if $n = 1$ according as g is orientation-preserving or orientation-reversing. In fact, we will use the same symbol $\pi_n(Y, y_0)$ for the group whose elements are classes of maps $I^n, \dot{I}^n \to Y, y_0$ or classes of maps $E^n, S^{n-1} \to Y, y_0$ or classes of maps $S^n, p_0 \to Y, y_0$, so that an element $\alpha \in \pi_n(Y, y_0)$ may be represented by a map of any of the given types. This notation will be justified, as we have said, in the next chapter, where it will be shown that there is, essentially, only one natural (1-1) correspondence between the classes of maps of different kinds.

For alternative descriptions of homotopy groups, see, for example, G. W. Whitehead, *Annals of Mathematics*, 51, 1950, 192–238; J. H. C. Whitehead, *Annals of Mathematics*, 42, 1941, 409–28.

3. The role of the base-point; operation of $\pi_1(Y, y_0)$ on $\pi_n(Y, y_0)$. We have defined a group $\pi_n(Y, y_0)$, which depends on an integer n, a space Y, and a point $y_0 \in Y$, called the base point. The group is written additively if $n > 1$, and multiplicatively if $n = 1$. We study now the effect on the group of the choice of y_0 as base point. We recall that Y is arcwise-connected.

LEMMA 3·1. *Let E^n be any n-element with boundary S^{n-1}. Then $E^n \times 0 \cup S^{n-1} \times I$ is a retract of $E^n \times I$.*

We take E^n as the Euclidean full-sphere $x_1^2 + \ldots + x_n^2 \leqslant 1$, and embed it in $(n+1)$-space. The points of E^n are vectors x of absolute value $\leqslant 1$. Let us write $|x|$ for the absolute value $\sqrt{(x_1^2 + \ldots + x_n^2)}$ of the vector $x = (x_1, \ldots, x_n)$. Then the retraction

$$\rho : E^n \times I \to E^n \times 0 \cup S^{n-1} \times I$$

is just the projection from the point $(0, \ldots, 0, 2)$ in R^{n+1}, namely,†

$$\rho(x,t) = \left(\frac{x}{|x|}, 2 - \frac{2-t}{|x|}\right), \quad |x| \geqslant 1 - \frac{t}{2},$$

$$= \left(\frac{2x}{2-t}, 0\right), \quad |x| \leqslant 1 - \frac{t}{2}.$$

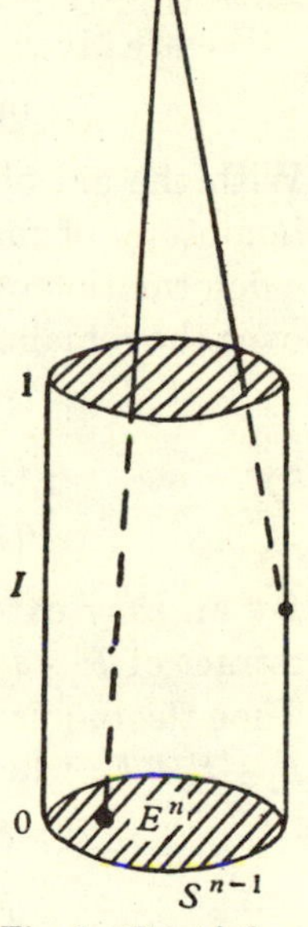

Fig. 1. The (deformation) retraction of $E^n \times I$ on to $E^n \times 0 \cup S^{n-1} \times I$.

It is clear that $E^n \times 0 \cup S^{n-1} \times I$ is, in fact, a deformation retract of $E^n \times I$. Now let C be a path from y_0 to y_1 in Y, i.e. a map $C : I \to Y$ with $C(0) = y_0$, $C(1) = y_1$. Let $\alpha_1 \in \pi_n(Y, y_1)$ be represented by $f_1 \in M_n(Y, y_1)$ and let

$$g' : I^n \times 0 \cup \dot{I}^n \times I \to Y$$

be given by $g'(x, 0) = f_1(x)$, $x \in I^n$, $g'(x,t) = C(1-t)$, $x \in \dot{I}^n$, $t \in I$. By the lemma there is a retraction $\rho : I^n \times I \to I^n \times 0 \cup \dot{I}^n \times I$. Then

† See fig. 1.

$g'\rho : I^n \times I \to Y$ is an extension of g' to $I^n \times I$. Let $f_0 \in M_n(Y, y_0)$ be given by $f_0(x) = g'\rho(x, 1)$, $x \in I^n$, and let $\alpha_0 = [f_0] \in \pi_n(Y, y_0)$. Let $[C]$ be the class of the map $C : I, 0, 1 \to Y, y_0, y_1$.

THEOREM 3·2. *The correspondence $\alpha_1 \to \alpha_0$ induces an isomorphism of $\pi_n(Y, y_1)$ on $\pi_n(Y, y_0)$ which depends only on $[C]$.*

Our first object is to show that if $F : I^n \times I \to Y$ is any map such that $F(x, 0) = f_1(x)$, $x \in I^n$, $F(x, t) = C'(1-t)$, $x \in \dot{I}^n$, $t \in I$, $C' \in [C]$, and if $f'_0 \in M_n(Y, y_0)$ is given by $f'_0(x) = F(x, 1)$, $x \in I^n$, then $[f'_0] = \alpha_0$. We first prove a special case of this, namely,

LEMMA 3·3. *Let $F : I^n \times I \to Y$ be a homotopy between*

$$f, f' \in M_n(Y, y_0)$$

such that $F(x, t) = \lambda(t)$, $x \in \dot{I}^n$, where $\lambda(t)$ is a nullhomotopic loop. Then $[f] = [f'] \in \pi_n(Y, y_0)$.

We are given a map $\mu : I \times I \to Y$ with

$$\mu(t, 0) = \lambda(t), \quad \mu(t, 1) = \mu(0, u) = \mu(1, u) = y_0.$$

With the aid of the map μ we replace the homotopy F by a homotopy of maps in $M_n(Y, y_0)$. Roughly speaking, we replace a deformation over one side of the (t, u)-square by a deformation over the remaining three sides. Precisely, define

$$j' : I^n \times I \times 0 \cup \dot{I}^n \times I \times I \to Y$$

by
$$j'(x, t, 0) = F(x, t), \quad x \in I^n, \ t \in I,$$
$$j'(x, t, u) = \mu(t, u), \quad x \in \dot{I}^n, \ (t, u) \in I \times I.$$

By an easy extension of lemma 3·1, $I^n \times I \times 0 \cup \dot{I}^n \times I \times I$ is a retract of $I^n \times I \times I$. Thus we may extend j' to $j : I^n \times I \times I \to Y$. Then the required homotopy between f and f' is the homotopy $F_1 + F_2 + F_3$, where $F_1, F_2, F_3 : I^n \times I, \dot{I}^n \times I \to Y, y_0$ are given by

$$\left.\begin{aligned} F_1(x, t) &= j(x, 0, t) \\ F_2(x, t) &= j(x, t, 1) \\ F_3(x, t) &= j(x, 1, 1-t) \end{aligned}\right\} \quad x \in I^n, \ t \in I.$$

This completes the proof of the lemma.

Returning to the theorem, let $k : I^n \times I \to Y$ be given by

$$\begin{aligned} k(x, t) &= g'\rho(x, 1-2t), \quad 0 \leqslant t \leqslant \tfrac{1}{2} \\ &= F(x, 2t-1), \quad \tfrac{1}{2} \leqslant t \leqslant 1. \end{aligned}$$

(Here $g'\rho$ is as in the preamble to the theorem and F as in the sentence following the statement of the theorem.) Then k is a homotopy between f_0 and f'_0 under which the point-image of $\dot{I}^n$ describes† CC'^{-1}. Since $C' \in [C]$, CC'^{-1} is a nullhomotopic loop, so that lemma 3·3 may be applied to show that $f'_0 \in [f_0] = \alpha_0$.

Now let $f'_1 \in \alpha_1$. Thus there exists a homotopy F_1 of maps in $M_n(Y, y_1)$ connecting f'_1 with f_1. The homotopy $F_1 + F$ connects f'_1 and f'_0 and the point-image of $\dot{I}^n$ describes $C_0 C'$, where C_0 is the constant loop (at y_1). Since $C_0 C' \in [C]$, it follows from what we have proved that α_0 depends only on α_1 and $[C]$. Let us call α_0 the image of α_1 under C^*. *Then C^* is a homomorphism.* For let $f_1, g_1, \in M_n(Y, y_1)$, represent $\alpha_1, \beta_1 \in \pi_n(Y, y_1)$ and let F, G be deformations of f_1, g_1 under which the point-image of $\dot{I}^n$ describes C^{-1}. Then $H, = F + G$, is defined since

$$F(1, t_2, \dots, t_n, t) = G(0, t_2, \dots, t_n, t) = C(1-t).$$

Moreover, $H(x, 0) = (f_1 + g_1)(x)$, $x \in I^n$, the map $f_0 \in M_n(Y, y_0)$ given by $f_0(x) = F(x, 1)$ represents $C^*(\alpha_1)$, the map $g_0 \in M_n(Y, y_0)$ given by $g_0(x) = G(x, 1)$ represents $C^*(\beta_1)$, and H is a deformation of $(f_1 + g_1)$ under which the point-image of $\dot{I}^n$ describes C^{-1}. Then, since $(f_1 + g_1)$ represents $\alpha_1 + \beta_1$, the map $h_0 \in M_n(Y, y_0)$ given by $h_0(x) = H(x, 1)$ represents $C^*(\alpha_1 + \beta_1)$. But $h_0 = f_0 + g_0$, so that h_0 represents $C^*(\alpha_1) + C^*(\beta_1)$.

Now let C_1 be a path from y_0 to y_1 and C_2 a path from y_1 to y_2. Then $C_1 C_2$ is a path from y_0 to y_2. If $f \in M_n(Y, y_2)$, then a deformation of f along C_2^{-1}, followed by a deformation along C_1^{-1}, constitutes a deformation of f along $(C_1 C_2)^{-1}$. Thus‡ $(C_1 C_2)^* = C_1^* C_2^*$. In particular, $(C^{-1}C)^* = (C^{-1})^* C^*$. Now $C^{-1}C$ is a nullhomotopic path (at y_1), so that $(C^{-1}C)^*$ is the identity automorphism of $\pi_n(Y, y_1)$. Similarly $C^*(C^{-1})^*$ is the identity automorphism of $\pi_n(Y, y_0)$. Thus $C^* : \pi_n(Y, y_1) \approx \pi_n(Y, y_0)$. This completes the proof of the theorem.

† $C_1 C_2$ is, by convention, the path C_1 *followed by* C_2.

‡ This is the reason why we take C^{-1} rather than C. If we had defined the operation C^* in terms of a deformation of $f \in M_n(Y, y_0)$ under which the point-image of $\dot{I}^n$ describes C_1 we would have $(C_1 C_2)^* = C^* C_1^*$ and $\pi_1(Y, y_0)$ anti-represented as a group of operators on $\pi_n(Y, y_0)$.

The abstract group of which $\pi_n(Y, y_0)$, $\pi_n(Y, y_1)$, ..., are isomorphic copies is called the nth homotopy group of Y and written $\pi_n(Y)$.

Now let y_1 coincide with y_0. Then the classes of closed paths based on y_0 are the elements of $\pi_1(Y, y_0)$. Thus $\pi_1(Y, y_0)$ acts as a group of operators on $\pi_n(Y, y_0)$. It is not difficult to see that $\pi_1(Y, y_0)$ acts on itself by inner automorphism, i.e. if

$$\alpha, \beta \in \pi_1(Y, y_0), \quad \text{then} \quad \beta(\alpha) = \beta\alpha\beta^{-1}.$$

We say that Y is *n-simple* if, for any two points y_1, y_2 and any two paths C_1, C_2 from y_1 to y_2, $C_1^* = C_2^*$. It may be shown that Y is n-simple if and only if, for some y_0, $\pi_1(Y, y_0)$ acts trivially on $\pi_n(Y, y_0)$. The necessity of the condition is obvious. Now let C be a path from y_0 to y_1. Then $CC_1C_2^{-1}C^{-1}$ is a closed path based on y_0, so that if $\pi_1(Y, y_0)$ acts trivially on $\pi_n(Y, y_0)$,

$$C^*C_1^*(C_2^*)^{-1}(C^*)^{-1} = 1 : \pi_n(Y, y_0) \approx \pi_n(Y, y_0).$$

Thus $C^*C_1^*(C_2^*)^{-1} = C^*$, $C_1^*(C_2^*)^{-1} = 1$, $C_1^* = C_2^*$ and Y is n-simple.

We see immediately that Y is 1-simple if and only if $\pi_1(Y)$ is Abelian and that Y is n-simple for all n if it is simply-connected (i.e. if $\pi_1(Y) = 0$).

Let us agree to identify elements of $\pi_n(Y, y_0)$ arising from maps $f: S^n, p_0 \to Y, y_0$ and $f': S'^n, p_0' \to Y, y_0$ if there exists an orientation-preserving homeomorphism $g: S^n, p_0 \to S'^n, p_0'$ such that $f = f'g$. With this convention, which will be justified in Chapter III, we may express the usefulness of the concept of n-simplicity in the following theorem:

THEOREM 3·4. *If Y is n-simple, a map $f: S^n \to Y$ determines a unique element of $\pi_n(Y, y_0)$ for each $y_0 \in Y$.*

For let $f(x_1) = y_1$, $x_1 \in S^n$, $y_1 \in Y$. Then f determines an element of $\pi_n(Y, y_1)$ and hence of $\pi_n(Y, y_0)$. Suppose that $f(x_2) = y_2$. Now there exists a rotation of S^n, say ρ, sending x_2 into x_1. Then if $f: S^n, x_1 \to Y, y_1$ represents $\alpha_1 \in \pi_n(Y, y_1)$, so does (by our convention) $f\rho: S^n, x_2 \to Y, y_1$. The image of the whole rotation of S^n under f is a homotopy of f under which the image of x_2 moves along a path C^{-1} from y_2 to y_1. Thus if $f: S^n, x_2 \to Y, y_2$

represents $\alpha_2 \in \pi_n(Y, y_2)$, $\alpha_1 = C^*(\alpha_2)$, so that the element of $\pi_n(Y, y_0)$ determined by f does not depend on the choice of base point in S^n.

We close this section by proving the theorem which gives the homotopy groups their topological significance.

THEOREM 3·5. *The homotopy groups are invariants of homotopy type.*

We first state a fundamental lemma. Its proof, however, is elementary and will only be sketched.

LEMMA 3·6. *Any map* $f: X, x_0 \to Y, y_0$ *induces a homomorphism* $f^*: \pi_n(X, x_0) \to \pi_n(Y, y_0)$. *If* $g: Y, y_0 \to Z, z_0$, *then* $(gf)^* = g^* f^*$.

Let $\alpha \in \pi_n(X, x_0)$ be represented by $h: I^n, \dot{I}^n \to X, x_0$. Define $f^*(\alpha)$ as the element of $\pi_n(Y, y_0)$ represented by $fh: I^n, \dot{I}^n \to Y, y_0$. The proof that f^* depends only on α, that it is a homomorphism, and that $(gf)^* = g^* f^*$ is elementary and is left to the reader.

We now revert to the theorem. Let X, Y be two spaces of the same homotopy type. In fact, let $f: X \to Y$, $g: Y \to X$ be maps such that $gf \sim 1: X \to X$, $fg \sim 1: Y \to Y$. Let $f(x_0) = y_0$, $g(y_0) = x_1$, $f(x_1) = y_1$. Then f induces

$$f^*: \pi_n(X, x_0) \to \pi_n(Y, y_0) \quad \text{and} \quad f^{**}: \pi_n(X, x_1) \to \pi_n(Y, y_1)$$

and g induces $g^*: \pi_n(Y, y_0) \to \pi_n(X, x_1)$. Let $\alpha \in \pi_n(X, x_0)$ be represented by $h: I^n, \dot{I}^n \to X, x_0$. Since

$$gf \sim 1: X \to X, \quad gfh \sim h: I^n \to X$$

and, moreover, throughout the homotopy $\dot{I}^n$ has a point-image which describes a path from x_0 to x_1. Thus $(gf)^*$, or $g^* f^*$, is an isomorphism of $\pi_n(X, x_0)$ on to $\pi_n(X, x_1)$. Similarly, $f^{**} g^*$ is an isomorphism of $\pi_n(Y, y_0)$ on to $\pi_n(Y, y_1)$. It follows from these two facts that g^* is a homomorphism of $\pi_n(Y, y_0)$ on to $\pi_n(X, x_1)$ and is an isomorphism. Thus $g^*: \pi_n(Y, y_0) \approx \pi_n(X, x_1)$, so that the homotopy groups of X and Y are isomorphic.

[*Note.* The contents of this section may all be 'reinterpreted' in terms of maps $S^n, p_0 \to Y, y_0$. We leave to the reader the definition of the operators and the 'translations' of the proofs of the theorems with this interpretation.]

4. The relative homotopy groups. Let Y_0 be a closed, arcwise-connected subspace of Y and let $y_0 \in Y_0$. Now I^{n-1} is the face of I^n given by $t_n = 0$; let J^{n-1} be the union of the remaining $(n-1)$-faces of I^n, and consider maps $f: I^n, I^{n-1}, J^{n-1} \to Y, Y_0, y_0$. We will call the totality of such maps $M_n(Y, Y_0, y_0)$ and the totality of homotopy classes of such maps $\pi_n(Y, Y_0, y_0)$. Now let $f, g \in M_n(Y, Y_0, y_0)$ and let $n \geqslant 2$. Then $f+g$ is defined and is in $M_n(Y, Y_0, y_0)$. Exactly as in § 1 of this chapter, we see that this addition in $M_n(Y, Y_0, y_0)$ induces an addition in $\pi_n(Y, Y_0, y_0)$. We will show that, with this addition, $\pi_n(Y, Y_0, y_0)$ is a group, and that it is Abelian if $n \geqslant 3$. It should be noted that the *relative homotopy group*, $\pi_n(Y, Y_0, y_0)$, reduces to the absolute homotopy group $\pi_n(Y, y_0)$ if $Y_0 = y_0$. However, we cannot in general give $\pi_1(Y, Y_0, y_0)$ a group structure, though it will sometimes prove useful to consider the set $\pi_1(Y, Y_0, y_0)$.

We now proceed much as in § 1 of this chapter. Let I_1^n be the subcube of I^n consisting of points $(t_1, \ldots, t_n)$ with $\lambda_i \leqslant t_i \leqslant \mu_i$, $i = 1, \ldots, n-1$, $0 \leqslant t_n \leqslant \mu_n$. We will only consider such subcubes in this section. Let $[f]$ stand for the homotopy class, in $\pi_n(Y, Y_0, y_0)$, containing the map $f \in M_n(Y, Y_0, y_0)$.

LEMMA 4·1. *Given $f \in M_n(Y, Y_0, y_0)$, there exists $f' \in [f]$ such that $f'(I^n - I_1^n) = y_0$.*

The proof is exactly as for lemma 1·1 (note that $\lambda_n = 0$).

We say that f' results from concentrating f on I_1^n.

Now let $f, g \in M_n(Y, Y_0, y_0)$, $n \geqslant 2$. We assume $f', \in [f]$, $g' \in [g]$ are concentrated on subcubes I_1^n, I_2^n, with disjoint interiors and I_1^n lying to the left of I_2^n. We define $k \in M_n(Y, Y_0, y_0)$ by

$$k \mid I_1^n = f', \quad k \mid I_2^n = g', \quad k(I^n - (I_1^n \cup I_2^n)) = y_0.$$

LEMMA 4·2. $k \in [f+g]$.

The proof is exactly as for lemma 1·2.

LEMMA 4·3. $[f+g] = [g+f]$ *if* $n > 2$.

Since $n > 2$, the proof of lemma 1·3 serves here also. Note, however, that, if $n = 2$, $\rho': I^2, \dot{I}^2 \to I^2, \dot{I}^2$ does not carry J^1 into J^1 and so does not lead to an admissible homotopy unless the map $f \in M_2(Y, Y_0, y_0)$ has the special property $f(\dot{I}^2) = y_0$.

THEOREM 4·4. *Under the operation of addition, the collection of classes, $\pi_n(Y, Y_0, y_0)$, is a group ($n \geqslant 2$). It is Abelian*† *if $n > 2$.*

The proof is exactly as for theorem 1·4.

Now I^n may be mapped homeomorphically on to the unit full-sphere E^n in such a way that I^{n-1} is mapped‡ on E^{n-1}_+ and J^{n-1} on E^{n-1}_-, and the point $x_0 = (0, 0, \ldots, 0) \in I^n$ is mapped on $p_0 = (1, 0, \ldots, 0) \in E^n$. Thus we may set up a (1-1) correspondence between classes of maps

$$I^n, \dot{I}^n, x_0 \to Y, Y_0, y_0$$

and classes of maps $E^n, S^{n-1}, p_0 \to Y, Y_0, y_0$. The relative homotopy groups were originally defined in terms of maps $E^n, S^{n-1}, p_0 \to Y, Y_0, y_0$ and we now proceed to describe them in such terms and to show the equivalence of the two definitions. In preparation for this, we prove two results.

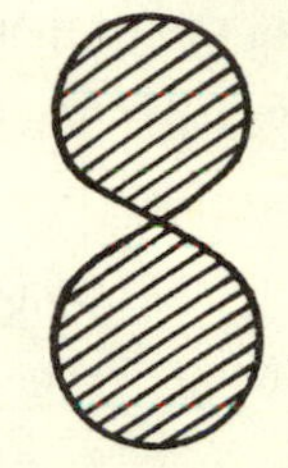

Fig. 2. If Y is the 'filled in' figure 8 and Y_0 its boundary, then $\pi_2(Y, Y_0) \approx \pi_1(Y_0)$, the free group on two generators.

Let $\phi_n : E^n_+, S^{n-1} \to S^n, p_0$ be defined as in § 2 of Chapter II, and let

$$\overline{\phi}_n : S^n, E^n_- \to S^n, p_0$$

be defined by $\overline{\phi}_n \mid E^n_+ = \phi_n$, $\overline{\phi}_n(E^n_-) = p_0$.

THEOREM 4·5. *The map $\overline{\phi}_n$ is homotopic to the identity. Precisely,*

$$\overline{\phi}_n \simeq 1 : S^n, E^n_-, p_0 \to S^n, E^n_-, p_0.$$

We recall that $\phi_n = rh$, where

$$h : E^n_+, S^{n-1} \to S^n, q_0 \quad \text{and} \quad r : S^n, q_0 \to S^n, p_0.$$

We extend h to $\overline{h} : S^n, E^n_- \to S^n, q_0$ by defining $\overline{h}(E^n_-) = q_0$, and then $\overline{\phi}_n = r\overline{h}$. Represent points of S^n by the pair (x, u), $x \in S^{n-1}$, $-1 \leqslant u \leqslant 1$, so that $(x, u) = (x_1\sqrt{1-u^2}, \ldots, x_n\sqrt{1-u^2}, u)$ if $x = (x_1, \ldots, x_n)$. Define

$$h_t : S^n \to S^n$$

by

$$\begin{aligned} h_t(x, u) &= (x, u - t + ut), \quad 0 \leqslant u \leqslant 1 \\ &= (x, u - t - ut), \quad -1 \leqslant u \leqslant 0. \end{aligned}$$

Then $h_0 = 1$, $h_1 = \overline{h}$. Define

$$r_t : S^n \to S^n$$

by

$$r_t(x_1, \ldots, x_{n+1}) = (x_1\sqrt{1-t^2} - x_{n+1}t, x_2, \ldots, x_n, x_1 t + x_{n+1}\sqrt{1-t^2})$$

† See fig. 2 for a pair (Y, Y_0) such that $\pi_2(Y, Y_0)$ is non-Abelian.

‡ As before, E^{n-1}_+ and E^{n-1}_- are the northern and southern hemispheres of the sphere S^{n-1} bounding E^n.

Then $r_0 = 1$, $r_1 = r$. Thus $\overline{\phi}_n = r_1 h_1$. The homotopy

$$r_t h_t : S^n \to S^n$$

has the properties stated in the theorem. For

$$r_t h_t(p_0) = r_t h_t(1, 0, \ldots, 0) = r_t(\sqrt{1-t^2}, 0, \ldots, 0, -t) = (1, 0, \ldots, 0) = p_0;$$

and if $u \leqslant 0$, $v = u - t - ut$, then

$$r_t h_t(x, u) = r_t(x, v) = (\ldots, t\sqrt{1-v^2}\, x_1 + v\sqrt{1-t^2}).$$

Since $-v = t - u + ut \geqslant t$, we have

$$-v\sqrt{1-t^2} \geqslant t\sqrt{1-v^2} \geqslant t\sqrt{1-v^2}\, x_1,$$

so that $r_t h_t(x, u) \in E^n_-$. This completes the proof of the theorem.

Let E^n_1, E^n_2 be the subsets of E^n given by $x_n \geqslant 0$, $x_n \leqslant 0$ respectively. We then have the

COROLLARY 4·6. *There exists a homotopy* $\omega \sim 1 : E^n, S^{n-1}, p_0 \to E^n, S^{n-1}, p_0$ *such that* $\omega(E^n_2) = p_0$.

The points of E^n may be represented as $\lambda x + (1-\lambda) p_0$, $0 \leqslant \lambda \leqslant 1$, $x \in S^{n-1}$; that is to say, the point $\lambda x + (1-\lambda) p_0$ has co-ordinates $(\lambda x_1 + 1 - \lambda, \lambda x_2, \ldots, \lambda x_n)$. Then†

$$\lambda x + (1-\lambda) p_0 \in E^n_1(E^n_2) \quad \text{if} \quad x \in E^{n-1}_+(E^{n-1}_-).$$

Let $g'_t = r_t h_t : S^{n-1}, E^{n-1}_-, p_0 \to S^{n-1}, E^{n-1}_-, p_0$. Then g'_t may be extended to

$$g_t : E^n, E^n_2, p_0 \to E^n, E^n_2, p_0$$

by defining $\quad g_t(\lambda x + (1-\lambda) p_0) = \lambda g'_t(x) + (1-\lambda) p_0.$

Since $g'_0 = 1$, it follows that $g_0 = 1$, and, if $x \in E^{n-1}_-$, then

$$\begin{aligned} g_1(\lambda x + (1-\lambda) p_0) &= \lambda g'_1(x) + (1-\lambda) p_0 \\ &= \lambda \overline{\phi}_{n-1}(x) + (1-\lambda) p_0 = \lambda p_0 + (1-\lambda) p_0 = p_0. \end{aligned}$$

Also $g_t(p_0) = p_0$ since $g'_t(p_0) = p_0$. This proves the corollary; we put $\omega = g_1$. We note for future application that we have actually proved that E^n_2 remains in E^n_2 throughout the homotopy between ω and the identity.

We now prove

THEOREM 4·7. *We may set up a* (1-1) *correspondence between classes of maps* $I^n, I^{n-1}, J^{n-1} \to Y, Y_0, y_0$ *and classes of maps* $E^n, S^{n-1}, p_0 \to Y, Y_0, y_0$.

† Note that $E^n_1 \cap S^{n-1} = E^{n-1}_+$, $E^n_2 \cap S^{n-1} = E^{n-1}_-$.

In fact, we will set up a (1-1) correspondence between classes of maps $E^n, E_+^{n-1}, E_-^{n-1} \to Y, Y_0, y_0$ and classes of maps

$$E^n, S^{n-1}, p_0 \to Y, Y_0, y_0.$$

Let $f: E^n, S^{n-1}, p_0 \to Y, Y_0, y_0$; then $f\omega \sim f: E^n, S^{n-1}, p_0 \to Y, Y_0, y_0$ and $f\omega(E_-^{n-1}) = f(p_0) = y_0$. Thus in each class of maps

$$E^n, S^{n-1}, p_0 \to Y, Y_0, y_0$$

there is a map $\quad E^n, E_+^{n-1}, E_-^{n-1} \to Y, Y_0, y_0.$

To establish that the correspondence of classes is (1-1), we have to prove that *if* $f, f': E^n, E_+^{n-1}, E_-^{n-1} \to Y, Y_0, y_0$ *and if*

$$f \sim f': E^n, S^{n-1}, p_0 \to Y, Y_0, y_0,$$

then $\quad f \sim f': E^n, E_+^{n-1}, E_-^{n-1} \to Y, Y_0, y_0.$

Now $f\omega \sim f: E^n, E_+^{n-1}, E_-^{n-1} \to Y, Y_0, y_0$, since

$$\omega \sim 1: E^n, E_-^{n-1}, p_0 \to E^n, E_-^{n-1}, p_0$$

and $f(E_-^{n-1}) = y_0$. Similarly $f'\omega \sim f': E^n, E_+^{n-1}, E_-^{n-1} \to Y, Y_0, y_0$. Also $f\omega \sim f'\omega: E^n, E_+^{n-1}, E_-^{n-1} \to Y, Y_0, y_0$, since $\omega(E_-^{n-1}) = p_0$. Thus $f \sim f\omega \sim f'\omega \sim f': E^n, E_+^{n-1}, E_-^{n-1} \to Y, Y_0, y_0$, and the theorem is proved.

Addition may be defined in the set of classes of maps

$$E^n, S^{n-1}, p_0 \to Y, Y_0, y_0$$

by using a fixed (1-1) correspondence between these classes and elements of $\pi_n(Y, Y_0, y_0)$ (as classes of maps $I^n, I^{n-1}, J^{n-1} \to Y, Y_0, y_0$) and the given addition of elements in $\pi_n(Y, Y_0, y_0)$. However, it is convenient to have the following direct definition. Given classes α, β, it follows readily from Corollary 4·6 that we may choose representative maps $f \in \alpha$, $g \in \beta$ such that $f(E_2^n) = g(E_1^n) = y_0$. We then define $\alpha + \beta$ as the class containing the map which coincides with f on E_1^n and with g on E_2^n. That $\alpha + \beta$, so defined, depends only on the classes α, β follows immediately from the following lemma, whose proof resembles closely an argument used in proving theorem 4·7.

LEMMA 4·8. *Let* $f_0 \sim f_1: E^n, S^{n-1}, p_0 \to Y, Y_0, y_0$ *with* $f_i(E_2^n) = y_0$, $i = 1, 2$. *Then there exists a homotopy* $f_t: E^n, S^{n-1}, p_0 \to Y, Y_0, y_0$ *with* $f_t(E_2^n) = y_0$.

For $\quad f_0 \sim f_0\omega \sim f_1\omega \sim f_1: E^n, S^{n-1}, E_2^n \to Y, Y_0, y_0.$

We leave to the reader the verification that we do in fact form the set of clases of maps $E^n, S^{n-1}, p_0 \to Y, Y_0, y_0$ into a group isomorphic with $\pi_n(Y, Y_0, y_0)$ as originally defined when we interpret $\alpha+\beta$ as above. We will use the symbol $\pi_n(Y, Y_0, y_0)$ to stand for the group defined in either of the two ways.

We have proved theorem 4·7 and lemma 4·8 by the use of the special homotopy $r_t h_t : S^n \to S^n$. It is possible to prove these propositions—as special cases of far more general propositions—by the use of the following fundamental theorem.

THEOREM 4·9. (The Homotopy Extension Theorem.) *Let K be a finite simplicial complex, and L a closed subcomplex. Let $f_0 : K \to Y$ and let $g_t : L \to Y$ be such that $g_0 = f_0 \mid L$. Then f_0 admits a homotopy $f_t : K \to Y$ such that $f_t \mid L = g_t$.*

We note first that lemma 3·1 leads immediately to a special case of this theorem, namely when K is an n-simplex and L its boundary. Now define $f_t(\sigma^0) = f_0(\sigma^0)$ for each vertex σ^0 of $K-L$, and extend g_t to the 1-simplexes of $K-L$ by lemma 3·1. The process may then be applied to the 2-simplexes of $K-L$, and so on until g_t has been extended to the whole of K.

A scrutiny of the proof justifies the following corollary.

COROLLARY 4·10. *Suppose, in addition, that, for some closed subcomplex M, $f_0(M) \subset Y_0$ and $g_t(L \cap M) \subset Y_0$. Then we may choose f_t so that $f_t(M) \subset Y_0$.*

Theorem 4·9 will be applied in subsequent chapters. The original proof of lemma 4·8 is to be found in J. H. C. Whitehead, On $\pi_r(V_{n,m})$ and sphere-bundles, *Proc. London Math. Soc.* 48 (1944), p. 281; it was based on the homotopy extension theorem.

Now suppose C is any path in Y_0 running from y_0 to y_1. Given a map $f_1 \in M_n(Y, Y_0, y_1)$ representing $\alpha_1 \in \pi_n(Y, Y_0, y_1)$, we will show how, as in § 3, we may deform f_1 along C^{-1} into a map $f_0 \in M_n(Y, Y_0, y_0)$. Let $F''' : I^{n-1} \times 0 \cup \dot{I}^{n-1} \times I \to Y_0$ be given by $F'''(x, 0) = f_1(x)$, $x \in I^{n-1}$, $F'''(x, t) = C(1-t)$, $x \in \dot{I}^{n-1}$, $t \in I$. By lemma 3·1, F''' may be extended to $F'' : I^{n-1} \times I \to Y_0$. Define $F' : I^n \times 0 \cup \dot{I}^n \times I \to Y$ by

$$F'(x, 0) = f_1(x), \quad x \in I^n, \quad F' \mid I^{n-1} \times I = F'', \quad F'(x, t) = C(1-t),$$
$$x \in J^{n-1}, \; t \in I.$$

Using lemma 3·1 again, we extend F' to $F: I^n \times I \to Y$. Then F is a homotopy of f_1 under which I^{n-1} stays in Y_0 and the point-image of J^{n-1} describes C^{-1}. Defining f_0 by $f_0(x) = F(x, 1)$, $x \in I^n$, we see that $f_0 \in M_n(Y, Y_0, y_0)$. Let $[f_0] = \alpha_0 \in \pi_n(Y, Y_0, y_0)$.

THEOREM 4·11. *The correspondence* $\alpha_1 \to \alpha_0$ *induces an isomorphism of* $\pi_n(Y, Y_0, y_1)$ *on* $\pi_n(Y, Y_0, y_0)$ *which depends only on the class of* $C: I, 0, 1 \to Y_0, y_0, y_1$.

The proof proceeds as for theorem 3·2. Owing to the close similarity of the two proofs, we will only give in detail the proof of the lemma which corresponds to lemma 3·3, and leave the reader to make the formal changes in the rest of the argument.

LEMMA 4·12. *Let* $F: I^n \times I, \dot{I}^n \times I \to Y, Y_0$ *be a homotopy between* $f, f' \in M_n(Y, Y_0, y_0)$ *such that* $F(x, t) = \lambda(t)$, $x \in J^{n-1}$, *where* $\lambda(t)$ *is a nullhomotopic loop (in* Y_0*). Then* $[f] = [f'] \in \pi_n(Y, Y_0, y_0)$.

We are given a map $\mu: I \times I \to Y_0$ with

$$\mu(t, 0) = \lambda(t), \quad \mu(t, 1) = \mu(0, u) = \mu(1, u) = y_0.$$

Define $j'''': I^{n-1} \times I \times 0 \cup \dot{I}^{n-1} \times I \times I \to Y_0$ by $j''''(x, t, 0) = F(x, t)$, $x \in I^{n-1}$, $t \in I$, $j''''(x, t, u) = \mu(t, u)$, $x \in \dot{I}^{n-1}$, $(t, u) \in I \times I$. Then, by an easy extension of lemma 3·1, j'''' may be extended to

$$j''': I^{n-1} \times I \times I \to Y_0.$$

Define $j'': \dot{I}^n \times I \times I \to Y_0$ by

$$j'' \mid I^{n-1} \times I \times I = j''', \quad j''(x, t, u) = \mu(t, u), \quad x \in J^{n-1},$$

and define $j': I^n \times I \times 0 \cup \dot{I}^n \times I \times I \to Y$ by

$$j' \mid \dot{I}^n \times I \times I = j'', \quad j'(x, t, 0) = F(x, t), \quad x \in I^n,\ t \in I.$$

Finally, we use lemma 3·1 again to extend j' to $j: I^n \times I \times I \to Y$. Then the required homotopy between f and f' is the homotopy $F_1 + F_2 + F_3$, where $F_1, F_2, F_3: I^n \times I, I^{n-1} \times I, J^{n-1} \times I \to Y, Y_0, y_0$ are given by

$$\left.\begin{aligned} F_1(x, t) &= j(x, 0, t) \\ F_2(x, t) &= j(x, t, 1) \\ F_3(x, t) &= j(x, 1, 1-t) \end{aligned}\right\} x \in I^n,\ t \in I.$$

This completes the proof of the lemma. The theorem may now be proved exactly as for theorem 3·2.

We have observed that $\pi_n(Y, y_0, y_0) = \pi_n(Y, y_0)$. However, the operation described above reduces to triviality if $Y_0 = y_0$, since C is a path in Y_0. Thus the operation described above cannot, strictly speaking, be called a generalization of the operation described for absolute homotopy groups. On the other hand, we have (taking $y_1 = y_0$) $\pi_1(Y_0, y_0)$ acting as a group of operators on $\pi_n(Y_0, y_0)$ and $\pi_n(Y, Y_0, y_0)$. Since any loop in Y_0 is, *a fortiori*, a loop in Y, we also have $\pi_1(Y_0, y_0)$ acting as a group of operators on $\pi_n(Y, y_0)$. This situation will be studied further in Chapter IV.†

The abstract group of which $\pi_n(Y, Y_0, y_0)$, $\pi_n(Y, Y_0, y_1)$, ..., are isomorphic copies is called the *nth relative homotopy group of the pair* (Y, Y_0).

The remaining definitions and theorems of § 3 may also be 'relativized'. Thus we say that the pair (Y, Y_0) is *n-simple* if, for any two points $y_1, y_2 \in Y_0$ and any two paths C_1, C_2 in Y_0 with end-points y_1, y_2, $C_1^* = C_2^* : \pi_n(Y, Y_0, y_2) \to \pi_n(Y, Y_0, y_1)$; it then follows as before that the pair (Y, Y_0) is n-simple if, and only if, for some $y_0 \in Y_0$, $\pi_1(Y_0, y_0)$ acts trivially on $\pi_n(Y, Y_0, y_0)$. If (Y, Y_0) is n-simple, then an element of $\pi_n(Y, Y_0)$ is uniquely determined‡ by a map $f : I^n, \dot{I}^n \to Y, Y_0$ (or a map of $f : E^n, S^{n-1} \to Y, Y_0$). Of course, if elements of $\pi_n(Y, Y_0, y_0)$ are represented by maps $f : E^n, S^{n-1}, p_0 \to Y, Y_0, y_0$, the operations of $\pi_1(Y_0, y_0)$ are defined in terms of deformations of f under which S^{n-1} stays in Y_0 and the image of p_0 describes a loop in the inverse class, $\in \pi_1(Y_0, y_0)$, to that of the operator.

We say that the pairs (Y, Y_0), (Z, Z_0) are of the same (relative) homotopy type if there exist maps $f : Y, Y_0 \to Z, Z_0$, $g : Z, Z_0 \to Y, Y_0$ such that $gf \sim 1 : Y, Y_0 \to Y, Y_0$, $fg \sim 1 : Z, Z_0 \to Z, Z_0$.

THEOREM 4·13. *The relative homotopy groups are invariants of (relative) homotopy type.*

This theorem follows formally from lemma 4·14 just as theorem 3·5 followed from lemma 3·6.

† We also make an algebraic study of the operation of $\pi_1(Y_0, y_0)$ on $\pi_2(Y, Y_0, y_0)$ in §4 of Chapter IV.

‡ Cf. theorem 3·4.

LEMMA 4·14. *Any map* $f: X, X_0, x_0 \to Y, Y_0, y_0$ *induces a homomorphism* $f^*: \pi_n(X, X_0, x_0) \to \pi_n(Y, Y_0, y_0)$. *If*

$$g: Y, Y_0, y_0 \to Z, Z_0, z_0, \quad \textit{then} \quad g^* f^* = (gf)^*.$$

Let $\alpha \in \pi_n(X, X_0, x_0)$ be represented by

$$h: I^n, I^{n-1}, J^{n-1} \to X, X_0, x_0.$$

Then fh is a map $I^n, I^{n-1}, J^{n-1} \to Y, Y_0, y_0$. Define $f^*(\alpha) = [fh]$. The proof of the lemma is now elementary.

CHAPTER III

THE CLASSICAL THEOREMS OF HOMOTOPY THEORY

(The standard reference throughout this chapter is to Lefschetz, *Introduction to Topology*, no. 11, Princeton Mathematical Series, Princeton University Press, 1949. We will abbreviate this, in references, to L.)

1. The simplicial approximation theorem. Let $K, L, \ldots$, stand for finite simplicial complexes and let $|K|, |L|, \ldots$, stand for their underlying polyhedra. We refer to L for the notions of a *barycentric subdivision* of a complex and the *barycentric co-ordinate system* of a simplex.

Given a transformation of the vertices of K into those of L, which sends vertices of a simplex of K into vertices of a simplex of L, it is possible to extend the transformation, by linearity within each simplex with respect to its barycentric co-ordinate system, to a map of $|K|$ into $|L|$. Such a map is called *simplicial* or *linear*. A simplicial map of $|K|$ into $|L|$, even if degenerate on some simplexes of K, induces a *chain-mapping* of the chains of K into the chains of L, that is to say, a collection of homomorphisms $h_n: C_n(K) \to C_n(L)$, $n = 0, 1, 2, \ldots$, of the integral† chains of K into those of L, such that

$$\delta_n h_n = h_{n-1} \delta_n, \; n = 1, 2, \ldots,$$

where δ_n is the (homology) boundary (in K and L). Where no confusion will arise, we will often use the same letter for a simplicial map and the induced chain-mapping.

THEOREM 1·1. (The Simplicial Approximation Theorem.) *Every mapping of $|K|$ into $|L|$ is homotopic to a simplicial mapping $|K_1| \to |L|$, where K_1 is a suitable subdivision of K.*

The proof is given in L, Chapter IV, § 4.

† We are only concerned with homology with integer coefficients.

THEOREM 1·2. $\pi_r(S^n) = 0$, $r < n$.

Let $\alpha \in \pi_r(S^n)$ be represented by a map $f: S^r, p_0 \to S^n, q_0$, where p_0, q_0 are chosen as base points in S^r, S^n. Give S^r, S^n simplicial structures. Then $f \sim f': S^r \to S^n$, where f' is simplicial with respect to some subdivision of the simplicial structure of S^r. Then $f'(S^r)$ contains no simplexes of dimension $\geqslant r$, so that $f'(S^r)$ is not the whole of S^n. Let $f'(p_0) = q_1$. Then $f'(S^r)$ is contractible, rel p_0, over S^n, so that f' represents the zero of $\pi_r(S^n, q_1)$. Thus α corresponds to 0 in some isomorphism between $\pi_r(S^n, q_0)$ and $\pi_r(S^n, q_1)$, whence $\alpha = 0$, and the theorem is proved.

It should be noted that the proof could be shortened a little by taking p_0 and q_0 as vertices of the complexes covering S^r and S^n and then keeping the image of p_0 at q_0 throughout the homotopy. Reference to the proof of theorem 1·1 shows that this is possible.

2. The Brouwer degree. Let $f: S_1^n \to S_2^n$ be a map of S_1^n into S_2^n. Let S_1^n, S_2^n be triangulated and let $f \sim f': S_1^n \to S_2^n$, where f' is simplicial with respect to some subdivision of the triangulation of S_1^n. Now, on any triangulation of an oriented n-sphere, there exists a fundamental† n-cycle (the positive generator of the cyclic infinite integral homology group of dimension n). Let γ_i^n, $i = 1, 2$, be the fundamental n-cycles on the given triangulations of S_i^n, subdivided suitably in the case of S_1^n. The chain-mapping f' induced by the simplicial map f' transforms cycles into cycles so that $f'(\gamma_1^n) = d\gamma_2^n$ for some integer d. Then d is said to be the (*Brouwer*) *degree* of the map f. Brouwer proved the following fundamental theorem:

THEOREM 2·1. *The degree d of the map $f: S_1^n \to S_2^n$ depends solely on the homotopy class of f.*

The proof is given in L, Chapter IV, § 5. We note the following consequences of the definition of degree and theorem 2·1.

(A) The degree of a‡ deformation $S^n \to S^n$ is 1 (since the identity map has degree 1).

† If $\sigma_1^n, \ldots, \sigma$ are the n-simplexes of the given triangulation, oriented in agreement with the given orientation of the n-sphere, then $\gamma_n = \sum_{i=1}^{t} \sigma_i^n$.

‡ A *deformation* $f: S^n \to S^n$ is a map such that $f \sim 1: S^n \to S^n$.

(B) The degree of an inessential map is 0 (since the constant map has degree 0).

(C) If $f: S_1^n \to S_2^n$ has degree d and $g: S_2^n \to S_3^n$ has degree d', then $gf: S_1^n \to S_3^n$ has degree dd'.

(D) The mapping $f \to d(f)$ induces a homomorphism† of $\pi_n(S^n)$ into the integers.

We prove (D). Certainly $f \to d(f)$ induces a single-valued mapping of $\pi_n(S^n)$ into Z_∞, the group of integers. Now let $f, g: S_1^n \to S^n$ represent $\alpha, \beta \in \pi_n(S^n)$. We may suppose f, g concentrated on disjoint subcomplexes P, Q of a certain triangulation of S_1^n, and, moreover, that they are simplicial with respect to a certain triangulation of S^n. Then if $h: S_1^n \to S^n$ is given by $h \mid P = f$, $h \mid Q = g$, $h(x) = q_0$, $x \in S_1^n - (P \cup Q)$, q_0 the base-point in S^n, then h represents $\alpha + \beta$ and h is simplicial. Let f, g, h stand also for the induced chain-mappings. Then

$$f(\gamma_1^n) = d(\alpha)\,\gamma^n, \quad g(\gamma_1^n) = d(\beta)\,\gamma^n$$

and, since P and Q are disjoint,

$$h(\gamma_1^n) = f(\gamma_1^n) + g(\gamma_1^n) = (d(\alpha) + d(\beta))\,\gamma_n,$$

where $d(\alpha) = d(f)$, $d(\beta) = d(g)$. Since $h(\gamma_1^n) = d(h)\,\gamma^n$ and h represents $\alpha + \beta$, (D) follows. If $n = 1$, we have to insist that σ_1^n precedes σ_2^n in the given orientation of S_1^n.

We may extend the notion of degree to maps $E_1^n, S_1^{n-1} \to E_2^n, S_2^{n-1}$, where $E_i^n, i = 1, 2$, is an n-element bounded by S_i^{n-1} and to maps $E_1^n, S_1^{n-1} \to S_2^n, p_2$. In the first case, we say that the degree of f: $E_1^n, S_1^{n-1} \to E_2^n, S_2^{n-1}$ is just the degree of $f \mid S_1^{n-1}: S_1^{n-1} \to S_2^{n-1}$. In the second case, we say that the degree of $g: E_1^n, S_1^{n-1} \to S_2^n, p_2$ is just the degree of $g\phi_n^{-1}: S_1^n \to S_2^n$, where ϕ_n is the map defined‡ in § 2 of Chapter II. From theorem 2·1 it follows that the degree is also an invariant of homotopy class of maps $E_1^n, S_1^{n-1} \to E_2^n, S_2^{n-1}$ and of maps $E_1^n, S_1^{n-1} \to S_2^n, p_2$.

† We drop the base-point from the symbol for a homotopy group where no confusion will arise.

‡ We identify the map $\phi_n: E_+^n \to S^n$ with the map $f: E^n \to S^n$ given by $f(x_1, \ldots, x_n) = \phi_n(x_1, \ldots, x_n, x_{n+1})$, where E^n is the vertical projection on the equatorial hyperplane of the northern hemisphere E_+^n. We will write in this section $\phi_n: E^n, S^{n-1} \to S^n, p_0$ or $\phi_n: E_1^n, S_1^{n-1} \to S_1^n, p_1$, where E_1^n is just a copy of E^n and (S_1^n, p_1) is a copy of (S^n, p_0).

The converse of Brouwer's theorem on degree is due to Hopf. The two theorems together assert

THEOREM 2·2. *Two maps* $f, g: S_1^n \to S_2^n$ *are homotopic if and only if they have the same degree.*

The proof of Hopf's theorem is given in L, Chapter IV, § 6. It proceeds by induction on n, proving simultaneously

THEOREM 2·3. *Two maps* $f, g: E_1^n, S_1^{n-1} \to E_2^n, S_2^{n-1}$, $n \geqslant 2$, *are homotopic if and only if they have the same degree.*

We will prove the easy consequence:

THEOREM 2·4. *Two maps* $f, g: E_1^n, S_1^{n-1} \to S_2^n, p_2$ *are homotopic if and only if they have the same degree.*

For let f, g have the same degree. Then $f\phi_n^{-1}$ and $g\phi_n^{-1}$ have the same degree, so that, by theorem 2·2, $f\phi_n^{-1} \sim g\phi_n^{-1}$, whence $f \sim g$. Conversely, let $f \sim g$. Then it was proved in the course of establishing theorem 2·1 of Chapter II that $f\phi_n^{-1} \sim g\phi_n^{-1}$. Thus by theorem 2·2 (or 2·1) $f\phi_n^{-1}$ and $g\phi_n^{-1}$ have the same degree, whence, by definition, f and g have the same degree.

It follows from (C) that any homeomorphism $S_1^n \to S_2^n$ has degree ± 1. We say that the homeomorphsm is *orientation-preserving* (*-reversing*) if its degree is $+1$ (-1). By theorem 2·2, any two orientation-preserving homeomorphisms of S_1^n on to S_2^n are homotopic. In particular, an orientation-preserving homeomorphism of S^n on to itself is homotopic to the identity map. Similarly, any two orientation-preserving homeomorphisms of E_1^n, S_1^{n-1} on to E_2^n, S_2^{n-1} are homotopic. Thus, as indicated in the previous chapter, the (1-1) correspondence set up between classes of maps $E^n, S^{n-1}, p_0 \to Y, Y_0, y_0$ and classes of maps $I^n, I^{n-1}, J^{n-1} \to Y, Y_0, y_0$ does not depend on the particular choice of orientation-preserving homeomorphism

$$I^n, \dot{I}^n \to E^n, S^{n-1};$$

and the algebraic structure of the groups $\pi_n(Y, Y_0)$ does not depend on whether an n-cube or a Euclidean n-element is used as counter-image.

Let $f: E_1^n, S_1^{n-1} \to S_2^n, p_2$ be any map such that $f \mid E_1^n - S_1^{n-1}$ is a homeomorphism on to $S_2^n - p_2$. Let $\phi_n: E_1^n, S_1^{n-1} \to S_1^n, p_1$ be the map of §2 of Chapter II. Then $f\phi_n^{-1}: S_1^n, p_1 \to S_2^n, p_2$ maps $S_1^n - p_1$ homeomorphically on to $S_2^n - p_2$ and is therefore a homeomorphism. We say that f is orientation-preserving if $f\phi_n^{-1}$ has degree $+1$, i.e. if f itself has degree $+1$. Then any two admissible† orientation-preserving maps $f, f': E_1^n, S_1^{n-1} \to S_2^n, p_2$ are homotopic. In particular, any map $f: E_1^n, S_1^{n-1} \to S_1^n, p_1$, of degree $+1$, is homotopic to ϕ_n, and so we have verified that the (1-1) correspondence set up between classes of maps

$$E^n, S^{n-1} \to Y, y_0$$

and classes of maps $S^n, p_0 \to Y, y_0$ does not depend on the particular choice of map ϕ_n; and the algebraic structure of the groups $\pi_n(Y)$ does not depend on whether an n-cube, a Euclidean n-element, or an n-sphere is used as counter-image.

It is now convenient to settle once and for all the question of the orientations of I^n and S^n. We do this by an inductive definition, first imbedding I^n in Euclidean n-space as the set of points $(x_1, \ldots, x_n)$ with $-1 \leqslant x_i \leqslant 1$, $i = 1, \ldots, n$. We orient I^1 as the directed line segment from -1 to 1. Let E^n be the n-element given by $x_1^2 + \ldots + x_n^2 \leqslant 1$. We assume S^{n-1} oriented. This induces an orientation of the pair E^n, S^{n-1}. Then S^n is oriented in such a way that ϕ_n is orientation-preserving and the pair $I^n, \dot{I}^n$ are oriented‡ in such a way that the radial projection

$$E^n, S^{n-1} \to I^n, \dot{I}^n$$

is orientation-preserving. This completes the inductive definition, since we may start with $n = 1$, where $S^0 = \dot{E}^1 = \dot{I}^1$. Having fixed orientations, we will agree to identify elements of $\pi_n(Y, y_0)$ represented by maps $f: E^n, S^{n-1} \to Y, y_0$ and $g: S^n, p_0 \to Y, y_0$ if $f\phi_n^{-1} \sim g: S^n, p_0 \to Y, y_0$.

THEOREM 2·5. *Let $f: E_+^n, S^{n-1} \to Y, y_0$ be extended to*

$$f': S^n, p_0 \to Y, y_0$$

by defining $f'(E_-^n) = y_0$. Then f and f' represent the same element of $\pi_n(Y, y_0)$.

† I.e. f and f' are both homeomorphisms of $E_1^n - S_1^{n-1}$ on to $S_1^n - p_1$. Of course, we might extend the use of the term 'orientation-preserving' to any map $E_1^n, S_1^{n-1} \to S_1^n, p_1$ of degree $+1$.

‡ An orientation of $I^n, \dot{I}^n$ is, more precisely, a generator of $H_n(I^n, \dot{I}^n)$.

Now the map $\phi_n : E^n_+, S^{n-1} \to S^n, p_0$ may be extended to $\overline{\phi}_n : S^n, p_0 \to S^n, p_0$ by defining $\overline{\phi}_n(E^n_-) = p_0$. We have proved that $\overline{\phi}_n \sim 1 : S^n, p_0 \to S^n, p_0$. The theorem is now proved by observing that

$$f\phi_n^{-1} = f'\overline{\phi}_n^{-1} : S^n, p_0 \to S^n, p_0;$$

for this implies that $f\phi_n^{-1} \sim f' : S^n, p_0 \to S^n, p_0$, and $f\phi_n^{-1}$, by our agreed identification, does represent the same element of $\pi_n(Y, y_0)$ as f.

Theorem 2·5 gives us a useful rule for obtaining maps of spheres representing given elements of $\pi_n(Y)$, when representative maps of n-elements are given.

Finally, we prove the following special consequence of theorem 2·2 and (D):

THEOREM 2·6. *$\pi_n(S^n)$ is cyclic infinite.*

We may consider the elements of $\pi_n(S^n)$ as classes of maps of S^n into itself. If $\alpha \in \pi_n(S^n)$ we write $d(\alpha)$ for the degree of any map in the class α. We have already seen that $\alpha \to d(\alpha)$ is a homomorphism of $\pi_n(S^n)$ into the (additive) group of integers. Since the identity map has degree 1, it is a homomorphism on to the group of integers. Now suppose $n \geqslant 2$. Since two maps

$$f, g : S^n, p_0 \to S^n, p_0$$

are homotopic as maps $f, g : S^n \to S^n$ only if they have the same degree, and since S^n is simply-connected, it follows that two maps $f, g : S^n, p_0 \to S^n, p_0$ represent the same element of $\pi_n(S^n)$ if they have the same degree, so that $\alpha \to d(\alpha)$ is an isomorphism.

The case $n = 1$ is slightly different. There are many proofs that the fundamental group of the circle is cyclic infinite. It is perhaps most convenient to observe that in Lefschetz's proof of theorem 2·2, case $n = 1$, given on p. 124 of L, the homotopy connecting a map f of S^1 into S^1 of degree ρ and a standard map g of degree ρ (multiplication by ρ). actually leaves p_0 fixed, if f is interpreted as a map $S^1, p_0 \to S^1, p_0$ and p_0 is the point '0 mod 1'. Thus two maps $S^1, p_0 \to S^1, p_0$ of the same degree are, in fact, homotopic as maps $S^1, p_0 \to S^1, p_0$.

COROLLARY 2·7. *S^n is r-simple for all r.*

This follows from theorem 1·2 if $n > 1$.

If $n = 1$, it follows from theorem 2·6 and corollary 1·6 of Chapter V.

3. The Hurewicz isomorphism theorem. We showed in the previous section that $\alpha \to d(\alpha)$ was a homomorphism (actually, an isomorphism) of $\pi_n(S^n)$ into (actually, on to) the integers. We may generalize this as follows. Let K be a finite connected simplicial complex and let σ^0 be a vertex of K. Let $\alpha \in \pi_n(|K|, \sigma^0)$ be represented by a simplicial map $f: S^n, p_0 \to |K|, \sigma^0$, S^n being given a certain triangulation. Then, if γ^n is the fundamental (integral) n-cycle on S^n, $f(\gamma^n)$ is an n-cycle of K, and it follows as in the case $K = S^n$ that the homology class of $f(\gamma^n)$ does not depend on the choice of f within its homotopy class. Thus '$\alpha \to$ homology class of $f(\gamma^n)$' induces a mapping

$$\omega: \pi_n(|K|, \sigma^0) \to H_n(K).$$

We show, exactly as in the special case, that ω is a homomorphism. On the other hand, it is not an isomorphism if $|K|$ is not n-simple; for, as may readily be seen, two elements of $\pi_n(|K|, \sigma^0)$ which are associated by an operator in $\pi_1(|K|, \sigma^0)$ have the same ω-image. However, in the original papers in which he first defined the concept of a homotopy group, Hurewicz proved the following fundamental theorem.

THEOREM 3·1. *If* $\pi_r(|K|) = 0$, $r = 1, \ldots, n-1$, $(n \geqslant 2)$, *then* $\omega: \pi_n(|K|) \approx H_n(K)$.

The proof of the theorem is given in L, Chapter V, § 5. It is worth noting that, without being explicitly stated, the following result† for simply-connected polyhedra $|K|$ is proved there:

THEOREM 3·2. *If $\alpha \in \pi_n(|K|)$ is such that $\omega(\alpha) = 0$, then α may be represented by a map $f: S^n \to |K^{n-1}|$.*

It might be said that a fundamental difficulty in homotopy theory consists in the fact that such a map is homologically trivial, but is not, in general, nullhomotopic. Theorem 3·1

† We use the notation K^r for the *r-section* of K, that is, the subcomplex of K consisting of simplexes whose dimensions do not exceed r.

follows from theorem 3·2 by the application of the following lemma, which is of general importance:

LEMMA 3·3. $\pi_r(Y)=0$, $r=1,\ldots,n-1$, *if and only if every map of a polyhedron of dimension* $\leqslant n-1$ *into* Y *is nullhomotopic.*

The sufficiency is obvious. Now let $f: K^{n-1}\to Y$ be a map of an $(n-1)$-dimensional complex† into Y. Since Y is arcwise-connected, we may suppose $f(K^0)=y_0$. Now assume that $f\sim f': K^{n-1}\to Y$ and $f'(K^r)=y_0$. If $r<n-1$, let σ^{r+1} be any $(r+1)$-simplex of K^{n-1}. Then $f' \mid \sigma^{r+1}$ is a map $\sigma^{r+1}, \dot\sigma^{r+1}\to Y, y_0$. Since $\pi_{r+1}(Y)=0$,
$$f' \mid \sigma^{r+1}\sim f'': \sigma^{r+1}, \dot\sigma^{r+1}\to Y, y_0$$
with $f''(\sigma^{r+1})=y_0$. We may proceed in this way for each $(r+1)$-simplex, getting a map which we will still call f'' such that $f' \mid K^{r+1}\sim f'': K^{r+1}, K^r\to Y, y_0$ and $f''(K^{r+1})=y_0$. It now follows from the homotopy extension theorem (theorem 4·9 of Chapter II) that f'' may be extended to $f'': K^{n-1}\to Y$ such that
$$f\sim f'\sim f'': K^{n-1}\to Y \quad \text{and} \quad f''(K^{r+1})=y_0.$$
We proceed in this way until we reach the required constant map.

The contents of this section may all be relativized. If L is a closed subcomplex of K and f a simplicial map of E^n, S^{n-1} into $|K|, |L|$, then the image under the chain-mapping f of the fundamental relative cycle of $E^n \bmod S^{n-1}$ is some relative cycle of $K \bmod L$ whose homology class depends only on the element of $\pi_n(|K|, |L|)$ represented by f. In this way we obtain a homomorphism $\omega: \pi_n(|K|, |L|)\to H_n(K, L)$. Theorem 3·1 then may be relativized as

THEOREM 3·4. *If* $\pi_r(|K|, |L|)=0$, $r=1,\ldots,n-1$, $(n\geqslant 2)$, *and if* $(|K|, |L|)$ *is* n*-simple, then* $\omega: \pi_n(|K|, |L|)\approx H_n(K, L)$.

The proof uses

THEOREM 3·5. *If* $\alpha\in\pi_n(|K|, |L|)$ *is such that* $\omega(\alpha)=0$, *and if* $(|K|, |L|)$ *is* n*-simple, then* α *may be represented by a map*
$$f: E^n, S^{n-1}\to |K^{n-1}| \cup |L|, |L|.$$

† It is very convenient to use the same symbol for a complex and its underlying space if no confusion is caused thereby. We demonstrate the utility of this 'abuse of language' in the proof of this lemma and of lemma 3·6.

Theorem 3·4 then follows from theorem 3·5 and

LEMMA 3·6. $\pi_r(Y, Y_0) = 0$, $r = 1, \ldots, n-1$, *if and only if every map of a polyhedron of dimension* $\leqslant n-1$ *into* Y *may be deformed into* Y_0.

The sufficiency is obvious. Now let $f: K^{n-1} \to Y$ be a map of an $(n-1)$-dimensional complex into Y. Since Y is arcwise-connected, we may suppose that $f(K^0) = y_0$. The vanishing of $\pi_1(Y, Y_0)$ means (conventionally) that any map $I^1, \dot{I}^1 \to Y, y_0$ may be deformed into a map $I_1, \dot{I}_1 \to Y_0, y_0$. Applying this to each 1-simplex, and then using the homotopy extension theorem, we find that f is homotopic to a map $f': K^{n-1} \to Y$ such that

$$f'(K^1) \subset Y_0 \quad \text{and} \quad f'(K^0) = y_0.$$

Now suppose that $f \sim f^{(r)}: K^{n-1} \to Y$ such that $f^{(r)}(K^r) \subset Y_0$ and $f^{(r)}(K^0) = y_0$. If $r < n-1$, let σ^{r+1} be any $(r+1)$-simplex of K^{n-1}. Then $f^{(r)} \mid \sigma^{r+1}$ is a map $f^{(r)} \mid \sigma^{r+1}: \sigma^{r+1}, \dot{\sigma}^{r+1}, \sigma^0 \to Y, Y_0, y_0$, where σ^0 is some vertex of σ^{r+1}. Since $\pi_{r+1}(Y, Y_0) = 0$, there exists a homotopy

$$f_t: \sigma^{r+1}, \dot{\sigma}^{r+1}, \sigma^0 \to Y, Y_0, y_0$$

with $f_0 = f^{(r)} \mid \sigma^{r+1}$, $f_1(\sigma^{r+1}) = y_0$. Replace σ^{r+1} by the Euclidean $(r+1)$-element E^{r+1}, given by $x_1^2 + \ldots + x_{r+1}^2 \leqslant 1$, and let S_λ^r be the 'latitude' $x_1^2 + \ldots + x_{r+1}^2 = \lambda$, $0 \leqslant \lambda \leqslant 1$. Define

$$f_t': E^{r+1}, S^r, p_0 \to Y, Y_0, y_0$$

by

$$\begin{aligned} f_t'(S_\lambda^r) &= f_t(S_{(1+t)\lambda}^r), \quad 0 \leqslant \lambda \leqslant \tfrac{1}{2}, \\ &= f_{2t(1-\lambda)}(S_{\lambda+t(1-\lambda)}^r), \quad \tfrac{1}{2} \leqslant \lambda \leqslant 1. \end{aligned}$$

Then $f_0' = f_0$, $f_t'(S^r) = f_t'(S_1^r) = f_0(S_1^r)$, and $f_1'(E^{r+1}) \subset Y_0$. Thus, returning to σ^{r+1}, we have proved that we may choose a homotopy of $f^{(r)} \mid \sigma^{r+1}$ which does not alter $f^{(r)} \mid \dot{\sigma}^{r+1}$ and which pushes the image of σ^{r+1} into Y_0. The argument now proceeds exactly as for lemma 3·3.

The proof of theorem 3·4 is very similar to that of theorem 3·1 (as given in L). However, there is a certain difference in the proof that ω is on to $H_n(K, L)$, which renders a statement of the proof of that part of the theorem desirable.

We will suppose the conditions of theorem 3·4 satisfied† and let $z^n = \Sigma \mu_i \sigma_i^n$ be a certain integral relative n-cycle of K mod L.

† The n-simplicity of the pair $(|K|, |L|)$ is not required for the *on to* property.

Now it follows from lemma 3·6 that we may deform the identity map $K \to K$, rel L, into a map $f: K \to K$ such that $f(K^{n-1}) \subset L$. We may suppose f to be simplicial as a map $K_1 \to K$, where K_1 is a subdivision of K.† Let z_1^n be the subdivision of z^n on K_1. Then $f(z_1^n)$ is a cycle of $K \bmod L$ in the same class as z^n. Moreover, $f(z_1^n)$ is a chain of the form $\Sigma \mu_i C_i^n$, where C_i^n is a chain whose boundary is in L, and which is the image under f of a subdivided simplex of K. Thus the homology classes of the chains C_i^n generate $H_n(K, L)$ and each class is clearly the image under ω of an element of $\pi_n(|K|, |L|)$.

The extension of the Hurewicz theorem to the relative case was first announced by Hurewicz himself.‡ He stated both the absolute and relative isomorphism theorem for general arcwise-connected Hausdorff spaces, using singular homology theory. It has been our object to avoid introducing singular homology, so we have not carried out these further generalizations.§

We close this chapter with an application of theorem 3·1.

THEOREM 3·7. *Let Y be the union of m n-spheres, $n > 1$, with a single common point. Then $\pi_n(Y)$ is the free Abelian group on m generators.*

For a slight adaptation of the argument of theorem 1·2 shows that $\pi_r(Y) = 0$, $r = 1, \ldots, n-1$, and $H_n(Y)$ is, of course, the direct sum of m cyclic infinite groups. The generators of $\pi_n(Y)$ are classes of maps $S^n \to S_i^n$ of degree 1, $i = 1, \ldots, m$.

† The use of singular homology avoids much of the difficulty.

‡ In a course of lectures.

§ For a not very detailed treatment of the general Hurewicz theorems, see S. T. Hu, 'An exposition of the relative homotopy theory', *Duke J. Math.* 14 (1947), 991–1033.

CHAPTER IV

THE EXACT HOMOTOPY SEQUENCE

1. Definition of the sequence. We first define the *boundary homomorphism* $d:\pi_n(Y,Y_0,y_0)\to\pi_{n-1}(Y_0,y_0)$, $n\geqslant 2$. Let

$$f:I^n,I^{n-1},J^{n-1}\to Y,Y_0,y_0$$

represent $\alpha\in\pi_n(Y,Y_0,y_0)$. Then $f\mid I^{n-1}:I^{n-1},\dot{I}^{n-1}\to Y_0,y_0$ represents an element $\beta\in\pi_{n-1}(Y_0,y_0)$. It is clear that β depends only on α and we write $\beta=d(\alpha)$. Then it is clear that d is a homomorphism. It is, moreover, an *operator* homomorphism with respect to operators from $\pi_1(Y_0,y_0)$, as may be verified immediately. Equivalently, we may represent $\alpha\in\pi_n(Y,Y_0,y_0)$ by a map $f':E^n,S^{n-1},p_0\to Y,Y_0,y_0$ and then $d(\alpha)$ will be represented by $f'\mid S^{n-1}$: S^{n-1}, $p_0\to Y_0$, y_0. We will not in future describe the parallel development of the theory in terms of maps of n-elements and n-spheres, but will leave it to the reader.

The identity maps $Y_0,y_0\to Y,y_0$, $Y,y_0,y_0\to Y,Y_0,y_0$ induce transformations

$$i:\pi_n(Y_0,y_0)\to\pi_n(Y,y_0),\quad j:\pi_n(Y,y_0)\to\pi_n(Y,Y_0,y_0)$$

which are obviously homomorphisms.

THEOREM 1·1. *i and j are operator homomorphisms with respect to operators from $\pi_1(Y_0,y_0)$.*

Let $f:I^n,\dot{I}^n\to Y_0,y_0$ represent $\alpha\in\pi_n(Y_0,y_0)$, and let

$$f\sim f':I^n\to Y_0$$

under a homotopy in which the point image of $\dot{I}^n$ describes a closed curve representing $\xi^{-1}\in\pi_1(Y_0,y_0)$. Then $f':I^n,\dot{I}^n\to Y_0,y_0$ represents $\xi(\alpha)$. The same map f', regarded now as a map $f':I^n,\dot{I}^n\to Y,y_0$, represents $i(\xi(\alpha))$. Now f, regarded as a map $f:I^n,\dot{I}^n\to Y,y_0$, represents $i(\alpha)$ and $f\sim f':I^n\to Y$ is a homotopy which, by definition of the operation of $\pi_1(Y_0,y_0)$ on $\pi_n(Y,y_0)$, transforms the element $i(\alpha)$ into $\xi(i(\alpha))$. Thus i is an operator homomorphism.

Similarly, let $\alpha \in \pi_n(Y, y_0)$ be represented by $f: I^n, \dot{I}^n \to Y, y_0$. Then a homotopy of f in which $\dot{I}^n$ always has a point image which describes a closed curve representing $\xi^{-1} \in \pi_1(Y_0, y_0)$ is, *a fortiori*, a homotopy in which I^{n-1} stays in Y_0 and J^{n-1} has a point-image describing a closed curve representing $\xi^{-1} \in \pi_1(Y_0, y_0)$. Thus $j(\xi(\alpha)) = \xi(j(\alpha))$.

We are therefore enabled to define the sequence of operator homomorphisms†

$$\to \dots \to \pi_{n+1}(Y, Y_0) \xrightarrow{d_{n+1}} \pi_n(Y_0) \xrightarrow{i_n} \pi_n(Y) \xrightarrow{j_n} \pi_n(Y, Y_0) \xrightarrow{d_n} \dots$$

$$\longrightarrow \pi_2(Y, Y_0) \xrightarrow{d_2} \pi_1(Y_0) \xrightarrow{i_1} \pi_1(Y) \xrightarrow{j_1} \pi_1(Y, Y_0) \longrightarrow 0.$$

It is called *the homotopy sequence of the pair* (Y, Y_0).

2. Proof of exactness. THEOREM 2·1. *The homotopy sequence of the pair* (Y, Y_0) *is exact, meaning that the kernel of each homomorphism is the image of the preceding one.*

We have to prove six assertions,

(1a) $ji\pi_n(Y_0) = 0$, (1b) $j^{-1}(0) \subset i\pi_n(Y_0)$,

(2a) $dj\pi_n(Y) = 0$, (2b) $d^{-1}(0) \subset j\pi_n(Y)$,

(3a) $id\pi_{n+1}(Y, Y_0) = 0$, (3b) $i^{-1}(0) \subset d\pi_{n+1}(Y, Y_0)$.

Proof of (1a). Let $\alpha \in \pi_n(Y_0)$ be represented by

$$f: I^n, \dot{I}^n \to Y_0, y_0.$$

Then $ji\alpha$ is represented by $f: I^n, I^{n-1}, J^{n-1} \to Y, Y_0, y_0$, where $f(I^n) \subset Y_0$. Since I^n is contractible‡ over itself to $x_0 = (0, 0, \dots, 0)$, it follows that $ji\alpha = 0$.

Proof of (1b). Let $\alpha \in j^{-1}(0)$. Then α is represented by a map $f_0: I^n, \dot{I}^n \to Y, y_0$ and there exists a homotopy

$$f_t: I^n, \dot{I}^n, x_0 \to Y, Y_0, y_0$$

such that $f_1(I^n) = y_0$. We seek a homotopy keeping $\dot{I}^n$ at y_0 and pushing I^n into Y_0. It is convenient for this proof to regard the

† We may define the transformation $j_1: \pi_1(Y) \to \pi_1(Y, Y_0)$, even though $\pi_1(Y, Y_0)$ is not a group. We write $\pi_1(Y, Y_0) \to 0$ because Y_0 is arcwise-connected, in order to fit in with theorem 2·1. Note that $\pi_1(Y, Y_0) = 0$ means that any path in Y with its end-points in Y_0 may be deformed, relative to its end-points, into a path in Y_0.

‡ Recall that we have shown that a homotopy of maps $I^n, \dot{I}^n, x_0 \to Y, Y_0, y_0$ does not affect the class of f.

co-ordinates of I^n as running from -1 to $+1$.† Let us write $\dot{I}^n_\lambda$ for the boundary of the cube

$$-\lambda \leqslant t_i \leqslant \lambda, \quad i=1, \ldots, n, \; 0 \leqslant \lambda \leqslant 1,$$

and define

$$f'_t(x) = f_t((1+t)\,x), \quad x \in \dot{I}^n_\lambda, \; 0 \leqslant \lambda \leqslant \tfrac{1}{2},$$

$$f'_t(x) = f_{2t(1-\lambda)}\left(\frac{\lambda + t(1-\lambda)}{\lambda}\, x\right), \quad x \in \dot{I}^n_\lambda, \; \tfrac{1}{2} \leqslant \lambda \leqslant 1.$$

The idea of the foregoing homotopy is to expand the 'half-cube' $\frac{1}{2} \leqslant t_i \leqslant \frac{1}{2}$, $i = 1, \ldots, n$, to cover I^n, at the same time carrying out the homotopy f_t, and to contract the remainder of I^n on to $\dot{I}^n$, at the same time 'slowing down' the homotopy f_t so that it is 'stationary' on $\dot{I}^n$. We see, in fact, that

$$f'_0 = f_0, \quad f'_t(\dot{I}^n) = f'_t(\dot{I}^n_1) = f_0(\dot{I}^n),$$

$$f'_1(\dot{I}^n_\lambda) = y_0, \; 0 \leqslant \lambda \leqslant \tfrac{1}{2}, \quad f'_1(\dot{I}^n_\lambda) \subset Y_0, \; \tfrac{1}{2} \leqslant \lambda \leqslant 1.$$

Thus f'_1 also represents α, and $\alpha \in i\pi_n(Y_0)$.

Proof of (2*a*). Let $\alpha \in \pi_n(Y)$ be represented by

$$f : I^n, \dot{I}^n \to Y, y_0.$$

Then $dj\alpha$ is represented by

$$f \mid I^{n-1} : I^{n-1}, \dot{I}^{n-1} \to Y_0, y_0;$$

but $f(I^{n-1}) = y_0$, so that $dj\alpha = 0$.

Proof of (2*b*). Let $\alpha \in d^{-1}(0)$. Then α is represented by

$$f : I^n, I^{n-1}, J^{n-1} \to Y, Y_0, y_0,$$

such that $f \mid I^{n-1}$ is homotopic, rel $\dot{I}^{n-1}$, to y_0. The homotopy yields a map $F' : I^n \times 0 \cup \dot{I}^n \times I \to Y$ $(F'(x,t) = y_0, x \in J^{n-1})$, which may be extended to

$$F : I^n \times I, I^{n-1} \times I, J^{n-1} \times I \to Y, Y_0, y_0.$$

Thus f is in the same class as a map which sends $\dot{I}^n$ to y_0, so that $\alpha \in j\pi_n(Y)$.

The proofs of (3*a*) and (3*b*) follow from the following lemma:

LEMMA 2·2. *A map $f : S^n, p_0 \to Y, y_0$ is nullhomotopic if and only if it has an extension $f' : E^{n+1} \to Y$.*

† This proof reinterprets that of lemma 3·6 of Chapter III in terms of maps of cubes. Not infrequently it is convenient to 'centre' the n-cube at $(0, \ldots, 0)$.

First let f have such an extension and let θ_t contract E^{n+1} on to p_0. Then $f'\theta_t \mid S^n$ is the required homotopy of f. Note that it is a homotopy rel p_0.

Now suppose there exists $f_t: S^n \to Y$ with $f_1 = f$, $f_0(S^n) = y_0$. The points of E^{n+1} may be represented as λx, $0 \leqslant \lambda \leqslant 1$, $x \in S^n$; define $f'(\lambda x) = f_\lambda(x)$.

Proof of (3*a*). Let $\alpha \in \pi_{n+1}(Y, Y_0)$ be represented by

$$f: E^{n+1}, S^n, p_0 \to Y, Y_0, y_0.$$

(It turns out to be more convenient for this proof to use maps of E^{n+1} rather than I^{n+1}.) Then $id\alpha$ is represented by a map $f \mid S^n: S^n, p_0 \to Y, y_0$ which has an extension to E^{n+1}. Thus $id\alpha = 0$.

Proof of (3*b*). Let $\alpha \in i^{-1}(0)$ be represented by $f: S^n \to Y_0$ and suppose f nullhomotopic in Y. Then f has an extension to $f': E^{n+1}, S^n \to Y, Y_0$, so that $\alpha \in d\pi_{n+1}(Y, Y_0)$.

3. Properties of the homotopy sequence. Let K be a (finite) simplicial complex, L a closed subcomplex, and σ^0 a vertex of L. Then a *homology sequence* may be defined,† namely,

$$\ldots \longrightarrow H_{n+1}(K, L) \xrightarrow{d'_{n+1}} H_n(L) \xrightarrow{i'_n} H_n(K) \xrightarrow{j'_n} H_n(K, L) \to \ldots.$$

In fact, i'_n is induced by regarding a cycle on L as a cycle on K, j'_n is induced by regarding a cycle on K as a relative cycle mod L, and d'_{n+1} is induced by taking the boundary of a relative cycle mod L; the boundary is, of course, a cycle of L. We leave to the reader the verification of the fact that in this way the appropriate homomorphisms are induced, and the proof of the following theorem, which is much more elementary than for the corresponding homotopy theorem (theorem 2·1).

THEOREM 3·1. *The homology sequence is exact.*

We have described in Chapter III certain standard homomorphisms, which we now call λ_n, μ_n, ν_n,

$$\lambda_n: \pi_n(|K|, \sigma^0) \to H_n(K), \quad \mu_n: \pi_n(|L|, \sigma^0) \to H_n(L),$$
$$\nu_n: \pi_n(|K|, |L|, \sigma^0) \to H_n(|K|, |L|).$$

† Here, as in Chapter III, we confine ourselves to integer coefficients, though theorem 3·1 and a generalization of the homomorphisms λ_n, μ_n, ν_n are valid for general coefficients.

It is then a matter of straightforward verification from the definitions that the commutativity relations,

$$i'_n \mu_n = \lambda_n i_n, \quad j'_n \lambda_n = \nu_n j_n, \quad d'_n \nu_n = \mu_{n-1} d_n, \tag{3·2}$$

hold between the homomorphisms of the homotopy and homology sequence. We say that (λ, μ, ν) is a *homomorphism* of the homotopy sequence into the homology sequence. It is called the *natural* homomorphism, and the homomorphisms λ_n, μ_n, ν_n are also called *natural*. We add the remark that theorem 3·1 and (3·2) may be generalized to general arcwise-connected Hausdorff spaces if singular homology is used.

We give now a second example of a homomorphism between sequences. Let f be a map of Y, Y_0, y_0 into Z, Z_0, z_0. Then, suppressing base-points, f induces homomorphisms

$$f_n^0 : \pi_n(Y) \to \pi_n(Z), \quad f_n^* : \pi_n(Y_0) \to \pi_n(Z_0),$$
$$\bar{f}_n : \pi_n(Y, Y_0) \to \pi_n(Z, Z_0).$$

It is then easy to verify the commutativity relations,

$$i_n f_n^0 = f_n^* i_n, \quad j_n f_n^* = \bar{f}_n j_n, \quad d_n \bar{f}_n = f_{n-1}^0 d_n, \tag{3·3}$$

between the homomorphisms of the homotopy sequences of (Y, Y_0) and (Z, Z_0), both sets of homomorphisms being written i, j, d in (3·3). Moreover, we may prove

THEOREM 3·4. *The homomorphism* $(f^*, f^0, \bar{f})$ *of the homotopy sequence of* (Y, Y_0) *into that of* (Z, Z_0) *is an operator homomorphism, meaning that, for all* $\xi \in \pi_1(Y_0)$, $\alpha \in \pi_n(Y_0)$, $\beta \in \pi_n(Y)$, $\gamma \in \pi_n(Y, Y_0)$,

$$\text{(i)}\ f_n^0(\xi\alpha) = f_1^0(\xi)\,(f_n^0(\alpha)), \quad \text{(ii)}\ f_n^*(\xi\beta) = f_1^0(\xi)\,(f_n^*(\beta)),$$
$$\text{(iii)}\ \bar{f}_n(\xi\gamma) = f_1^0(\xi)\,(\bar{f}_n(\gamma)).$$

To prove (i), let $\alpha \in \pi_n(Y_0, y_0)$ be represented by $g : I^n, \dot{I}^n \to Y_0, y_0$ and let $G : I^n \times I \to Y_0$ be a homotopy of g under which the point-image of $\dot{I}^n$ describes a loop in the class ξ^{-1}. Then

$$fG : I^n \times I \to Z_0$$

is a homotopy of fg under which the point-image of $\dot{I}^n$ describes a loop in the class $f_1^0(\xi^{-1}) = (f_1^0(\xi))^{-1}$. Thus if $g' : I^n, \dot{I}^n \to Y_0, y_0$ is given by $g'(x) = G(x, 1)$, $x \in I^n$, then g' represents $\xi\alpha$ and fg' represents $f_1^0(\xi)\,(f_n^0(\alpha))$. Thus $f_n^0(\xi\alpha) = f_1^0(\xi)\,(f_n^0(\alpha))$. The proofs of (ii) and (iii) are similar.

If $f_n^*, f_n^0, \bar{f}_n$ are isomorphisms for all n, we describe $(f^*, f^0, \bar{f})$ as an isomorphism. It is clear that the homotopy sequences of (Y, Y_0) and (Z, Z_0) are isomorphic if (Y, Y_0) and (Z, Z_0) are of the same homotopy type. However, it does not necessarily follow that an isomorphism between homotopy sequences can be realized geometrically, in the sense of being induced by some map $f: Y, Y_0 \to Z, Z_0$. For 'good' spaces,† a map f which induces an isomorphism between the homotopy sequences may be shown to be a homotopy equivalence.

4. The group $\pi_2(Y, Y_0)$. In this section we study the 'bottom end' of the homotopy sequence, which presents an interesting algebraical situation. The results of this section will not be needed in the sequel, except when considering a special case of theorem 5·1, and even then a reference to corollary 4·2 could be avoided.

Consider then the sequence

$$\ldots \to \pi_2(Y) \xrightarrow{j} \pi_2(Y, Y_0) \xrightarrow{d} \pi_1(Y_0) \xrightarrow{i} \pi_1(Y) \to \pi_1(Y, Y_0) \to 0.$$

THEOREM 4·1. *$\pi_2(Y, Y_0)$ is a crossed $(\pi_1(Y_0), d)$-module. That is to say, it admits $\pi_1(Y_0)$ as a group of operators and there is a homomorphism $d: \pi_2(Y, Y_0) \to \pi_1(Y_0)$, such that*

$$\text{(i)}\quad d(\xi\alpha) = \xi(d\alpha)\,\xi^{-1},$$

$$\text{(ii)}\quad (d\alpha)(\beta) = \alpha + \beta - \alpha, \quad \alpha, \beta \in \pi_2(Y, Y_0), \quad \xi \in \pi_1(Y_0).$$

It remains to verify (i) and (ii).

To‡ prove (i), let $f \in M_2(Y, Y_0, y_0)$ represent α and let

$$F: I^2 \times I, I^1 \times I \to Y, Y_0$$

be a homotopy of f under which the point-image of J^1 describes a loop in ξ^{-1}. Since $F \mid I^1 \times I$ is a map of the whole square into

† Certainly for finite simplicial complexes, and even for the more general complexes of Chapter VII.

‡ We mentioned in Chapter II that if $\xi, \eta \in \pi_1(Y_0)$, then $\xi(\eta) = \xi\eta\xi^{-1}$. Since d is an operator homomorphism, $d(\xi\alpha) = \xi(d\alpha)$, so that (i) is just a special case of $\xi(\eta) = \xi\eta\xi^{-1}$. This general result is proved in the same way as (i).

Y_0, the map $F \mid (I^1 \times I)^{\cdot}$ represents the zero of the fundamental group of Y_0. Also the map $f' : I^1, I^1 \to Y_0, y_0$ given by

$$f'(t_1) = F(t_1, 1), \quad t_1 \in I^1$$

represents $d(\xi\alpha)$. Starting from $(0, 1)$ and going round the square $I^1 \times I$ anti-clockwise, we get $\xi \,.\, d\alpha \,.\, \xi^{-1} \,.\, (d(\xi\alpha))^{-1} = 1$ or

$$d(\xi\alpha) = \xi d\alpha \xi^{-1}.$$

To prove (ii), it is more convenient to represent α, β by maps $f, g : E^2, S^1, p_0 \to Y, Y_0, y_0$ such that $f(E^2_2) = g(E^2_1) = y_0$. Then $\alpha + \beta$ is represented by that map h which agrees with f on E^2_1 and with g on E^2_2. Let ρ_t, $0 \leqslant t \leqslant 1$, be a rotation of E^2 through an angle πt in the anti-clockwise sense. Let f_t, g_t, h_t be $f\rho_t, g\rho_t, h\rho_t$ respectively. Now under ρ_t, p_0 describes E^1_+ (the northern hemisphere of S^1). Thus g_t is a homotopy $g_t : E^2, S^1, p_0 \to Y, Y_0, y_0$, so that g_1 represents β. On the other hand, the image of p_0 under both f_t and h_t is a loop in Y_0 representing $d\alpha$. Thus f_1 represents $(d\alpha)^{-1}(\alpha)$ and h_1 represents $(d\alpha)^{-1}(\alpha + \beta)$. Now ρ_1 interchanges E^2_1 and E^2_2, so that $f_1(E^2_1) = g_1(E^2_2) = y_0$ and h_1 agrees with g_1 on E^2_1 and with f_1 on E^2_2. Thus h_1 represents $\beta + (d\alpha)^{-1}(\alpha)$, whence

$$(d\alpha)^{-1}(\alpha + \beta) = \beta + (d\alpha)^{-1}(\alpha), \quad \text{or} \quad \alpha + \beta - \alpha = (d\alpha)(\beta).$$

This proves (ii). Note that, in fact, $(d\alpha)^{-1}(\alpha) = \alpha$.

COROLLARY 4·2. *$j\pi_2(Y)$ is contained in the centre of $\pi_2(Y, Y_0)$.*

Let $\alpha \in j\pi_2(Y)$, $\beta \in \pi_2(Y, Y_0)$. Then, by theorem 2·1, $d\alpha = 1$. Thus, by (ii) of theorem 4·1, $\alpha + \beta - \alpha = \beta$ or $\alpha + \beta = \beta + \alpha$.

COROLLARY 4·3. *$i\pi_1(Y_0)$ acts as a group of operators on $j\pi_2(Y)$.*

Let $\alpha \in j\pi_2(Y)$, $\xi \in \pi_1(Y_0)$. By (i) of theorem 4·1, $d(\xi\alpha) = \xi\, 1\, \xi^{-1} = 1$, so that $\xi\alpha \in j\pi_2(Y)$. If $\xi = d\beta$, $\beta \in \pi_2(Y, Y_0)$, then $\xi\alpha = \beta + \alpha - \beta = \alpha$, since α is in the centre of $\pi_2(Y, Y_0)$. Thus $d\pi_2(Y, Y_0)$ acts trivially on $j\pi_2(Y)$. Thus the group $\pi_1(Y_0)/d\pi_2(Y, Y_0)$ acts on $j\pi_2(Y)$. But, by the exactness of the homotopy sequence, i induces an isomorphism $\pi_1(Y_0)/d\pi_2(Y, Y_0) \approx i\pi_1(Y_0)$. This establishes the corollary.

If Y is a 2-dimensional (connected) simplicial complex and Y_0 its 1-dimensional section, then it may be shown that

$$i\pi_1(Y_0) = \pi_1(Y) \quad \text{and} \quad \pi_2(Y_0) = 0,$$

so that j is an isomorphism. The operation described in corollary 4·3 then coincides with the ordinary operation of $\pi_1(Y)$ on $\pi_2(Y)$. The theory of crossed modules has been studied by J. H. C. Whitehead, *Bull. Amer. Math. Soc.* 55 (1949), 453–96, where he showed that, in this particular case, $\pi_2(Y, Y_0)$ is, in a certain sense, a 'free' crossed module.†

5. Special cases. THEOREM 5·1. *If Y_0 is deformable rel y_0 over Y to y_0, then*

$$\pi_n(Y, Y_0) \approx \pi_n(Y) + \pi_{n-1}(Y_0), \quad n \geqslant 2.$$

Let $h: Y_0 \times I \to Y$ be a map such that

$$h(p, 0) = p, \; p \in Y_0, \quad h(y_0, t) = h(p, 1) = y_0, \; p \in Y_0, \; t \in I.$$

Given $f: I^{n-1}, \dot{I}^{n-1} \to Y_0, y_0$, define $f': I^n, I^{n-1}, J^{n-1} \to Y, Y_0, y_0$ by $f'(t_1, \ldots, t_n) = h(f(t_1, \ldots, t_{n-1}), t_n)$. Then $f \to f^*$ clearly induces a homomorphism $\mu: \pi_{n-1}(Y_0) \to \pi_n(Y, Y_0)$. Consider the sequence

$$\pi_{n+1}(Y, Y_0) \xrightarrow{d_{n+1}} \pi_n(Y_0) \xrightarrow{i} \pi_n(Y) \xrightarrow{j} \pi_n(Y, Y_0) \xrightarrow{d_n} \pi_{n-1}(Y_0).$$

Then‡ $d\mu = 1$, so that, if $n > 2$,

$$\pi_n(Y, Y_0) = d^{-1}(0) + \mu\pi_{n-1}(Y_0) = j\pi_n(Y) + \mu\pi_{n-1}(Y_0).$$

Now μ is isomorphic. Also d_{n+1} is on to $\pi_n(Y_0)$, so that, by exactness, i is on to 0, and j is isomorphic. If $n = 2$, it again follows that any element in $\pi_n(Y, Y_0)$ is uniquely expressible as

$$x + y, \quad x \in j\pi_n(Y), \quad y \in \mu\pi_{n-1}(Y_0)$$

and j and μ are isomorphic. That x commutes with y follows from the fact that $j\pi_2(Y)$ is in the centre of $\pi_2(Y, Y_0)$. (This theorem applies when Y is a sphere and Y_0 a proper closed subset.)

THEOREM 5·2. *If Y_0 is a retract of Y, then*

$$\pi_n(Y) \approx \pi_n(Y_0) + \pi_n(Y, Y_0), \quad n \geqslant 2.$$

Let $k: Y \to Y_0$ be the retraction, so that $k \mid Y_0 = 1$. Thus, if i is the identity map $Y_0 \to Y$, $ki = 1$, so that $k^* i = 1$, where

$$k^*: \pi_n(Y) \to \pi_n(Y_0)$$

† Further contributions have been made by W. H. Cockcroft (*Quart. J. Math.* 2 (1951), 123–34, 159–60) and by Anne Cobbe (ibid. pp. 269–85).

‡ Here, and subsequently, we drop the dimensional suffix on the homomorphisms i, j, d where no confusion will arise.

is induced by k, and i is the injection $\pi_n(Y_0) \to \pi_n(Y)$ induced by i. Thus $\pi_n(Y) = i\pi_n(Y_0) + k^{*-1}(0)$, $n \geqslant 2$, and i is isomorphic, $n \geqslant 1$. Returning to the homotopy sequence, we see that d_{n+1} is on to 0, $n \geqslant 1$, so that j_n is on to $\pi_n(Y, Y_0)$, $n \geqslant 2$. Since the kernel of j is $i\pi_n(Y_0)$, $j \mid k^{*-1}(0)$ is an isomorphism of $k^{*-1}(0)$ on to $\pi_n(Y, Y_0)$, so that $\pi_n(Y) \approx \pi_n(Y_0) + \pi_n(Y, Y_0)$, $n \geqslant 2$.

THEOREM 5·3. *If Y may be deformed, rel y_0, into Y_0, then*

$$\pi_n(Y_0) \approx \pi_n(Y) + \pi_{n+1}(Y, Y_0), \quad n \geqslant 2.$$

Let $h_t: Y, y_0 \to Y, y_0$ be the deformation, let $f: Y, y_0 \to Y_0, y_0$ be given by $f(p) = h_1(p)$, $p \in Y$, and let f induce $f^*: \pi_n(Y) \to \pi_n(Y_0)$. Then $if \sim 1: Y, y_0 \to Y, y_0$, so that $if^* = 1: \pi_n(Y, y_0) \approx \pi_n(Y, y_0)$. Thus $\pi_n(Y_0) = f^*\pi_n(Y) + i^{-1}(0)$, $n \geqslant 2$, f^* is isomorphic, and i is on to $\pi_n(Y)$, $n \geqslant 1$. Returning to the homotopy sequence, since i is on to, j is on to 0, and d is isomorphic, so that

$$\pi_n(Y_0) = f^*\pi_n(Y) + d\pi_{n+1}(Y, Y_0),$$

and f^* and d are isomorphic.†

Not only theorems 5·1, 5·2 and 5·3, but also the precise way in which the subgroups are imbedded as direct summands in the appropriate groups, are important. We also note as a corollary to theorem 5·2 that if Y_0 is a retract of Y, then $\pi_2(Y, Y_0)$ is Abelian.

6. The homotopy groups of the union of two spaces.

Let Y, Y' be two arcwise-connected Hausdorff spaces with a single common point y_0, and let $Y \cup Y'$ be the space consisting of their union.‡ We identify $Y \cup Y'$ with the subset

$$Y \times y_0' \cup y_0 \times Y'$$

of $Y \times Y'$, where (y_0, y_0') is the common point of $Y \times y_0'$, $y_0 \times Y$.

THEOREM 6·1. $\pi_n(Y \times Y') \approx \pi_n(Y) + \pi_n(Y')$.

THEOREM 6·2.

$$\pi_n(Y \cup Y') \approx \pi_n(Y) + \pi_n(Y') + \pi_{n+1}(Y \times Y', Y \cup Y'), \quad n \geqslant 2.$$

† The reader will note the similarity in the technique of proof of 5·1 and 5·3.
‡ That is to say, Y and Y' retain their topologies as closed subsets of $Y \cup Y'$

(It may be shown that, if Y, Y' are polyhedra, then $\pi_1(Y \cup Y')$ is the free product of $\pi_1(Y)$ and $\pi_1(Y')$.)

Let $m: Y \times Y' \to Y$ be the projection defined by $m(y, y') = y$. and let $i: Y \to Y \times Y'$ be the identity map $i(y) = (y, y_0')$. Define m', i' similarly and let m, m', i, i' induce

$$\mu: \pi_n(Y \times Y') \to \pi_n(Y), \quad \mu': \pi_n(Y \times Y') \to \pi_n(Y'),$$
$$i: \pi_n(Y) \to \pi_n(Y \times Y'), \quad i': \pi_n(Y') \to \pi_n(Y \times Y').$$

Then it is clear that $\mu i = \mu' i' = 1$, $\mu i' = \mu' i = 0$ (the 'zero' homomorphism.

Let $\eta: \pi_n(Y \times Y') \to \pi_n(Y) + \pi_n(Y')$ be the homomorphism given by

$$\eta(\alpha) = \mu(\alpha) + \mu'(\alpha).$$

We show that η is an isomorphism. Since $\mu i = 1$, $\mu' i' = 1$, $\mu i' = \mu' i = 0$, it follows that

$$\eta(i\beta + i'\beta') = \beta + \beta', \quad \beta \in \pi_n(Y), \beta' \in \pi_n(Y').$$

Thus η is on to. Now let $\eta(\alpha) = 0$, and let $f: I^n, \dot{I}^n \to Y \times Y', (y_0, y_0')$ represent α. Let $f(x) = (g(x), g'(x))$, $x \in I^n$, so that $g: I^n, \dot{I}^n \to Y, y_0$ represents $\mu(\alpha)$ and $g': I^n, \dot{I}^n \to Y', y_0$ represents $\mu'(\alpha)$. Since $\mu(\alpha) = \mu'(\alpha) = 0$, there exist homotopies

$$g_t: I^n, \dot{I}^n \to Y, y_0, \quad g_t': I^n, \dot{I}^n \to Y', y_0'$$

with $g_0 = g$, $g_1(I^n) = y_0$, $g_0' = g'$, $g_1'(I^n) = y_0'$. The homotopy

$$f_t(x) = (g_t(x), g_t'(x))$$

then deforms f to the constant map and theorem 6·1 is proved.

Let $\bar{\imath}: \pi_n(Y) \to \pi_n(Y \cup Y')$, $\bar{\imath}': \pi_n(Y') \to \pi_n(Y \cup Y')$,

$$i^*: \pi_n(Y \cup Y') \to \pi_n(Y \times Y')$$

be the injections, and define

$$\theta: \pi_n(Y) + \pi_n(Y') \to \pi_n(Y \times Y'),$$
$$\bar{\eta}: \pi_n(Y \times Y') \to \pi_n(Y \cup Y')$$

by $\theta(\beta + \beta') = i\beta + i'\beta'$, $\bar{\eta}(\alpha) = \bar{\imath}\mu(\alpha) + \bar{\imath}'\mu'(\alpha)$, $\beta \in \pi_n(Y)$, $\beta' \in \pi_n(Y')$, $\alpha \in \pi_n(Y \times Y')$. Then $\eta\theta(\beta + \beta') = \eta(i\beta + i'\beta') = \beta + \beta'$, so that $\eta\theta = 1$, whence, η being an isomorphism, $\theta\eta = 1$. It follows that $i^*\bar{\eta}(\alpha) = i^*(\bar{\imath}\mu(\alpha) + \bar{\imath}'\mu'(\alpha)) = i\mu(\alpha) + i'\mu'(\alpha) = \theta\eta(\alpha) = \alpha$. Thus $\bar{\eta}$ is a right inverse of i^*. Also $\bar{\eta}i\beta = \bar{\imath}\mu i\beta + \bar{\imath}'\mu' i\beta = \bar{\imath}\beta$, so that $\bar{\eta}i = \bar{\imath}$, $\bar{\eta}i' = \bar{\imath}'$, and

$$\bar{\eta}\pi_n(Y \times Y') = \bar{\eta}i\pi_n(Y) + \bar{\eta}i'\pi_n(Y') = \bar{\imath}\pi_n(Y) + \bar{\imath}'\pi_n(Y'),$$

whence
$$\pi_n(Y \cup Y') = \bar{\eta}\pi_n(Y \times Y') + i^{*-1}(0)$$
$$= \bar{\imath}\pi_n(Y) + \bar{\imath}'\pi_n(Y') + i^{*-1}(0).$$
Since $Y \cup Y'$ may be projected on to Y or Y', $\bar{\imath}$ and $\bar{\imath}'$ are clearly univalent.

Consider the homotopy sequence of the pair $Y \times Y'$, $Y \cup Y'$
$$\ldots \rightarrow \pi_{n+1}(Y \times Y', Y \cup Y') \xrightarrow{d} \pi_n(Y \cup Y')$$
$$\xrightarrow{i^*} \pi_n(Y \times Y') \xrightarrow{j} \pi_n(Y \times Y', Y \cup Y') \xrightarrow{d_n} \pi_{n-1}(Y \cup Y') \rightarrow \ldots.$$
Then $i^{*-1}(0) = d\pi_{n+1}(Y \times Y', Y \cup Y')$. Since i^* is on to,† $n \geqslant 2$, j is on to 0, $n \geqslant 2$, so that d_n is isomorphic, $n \geqslant 2$. Thus
$$\pi_n(Y \cup Y') = \bar{\imath}\pi_n(Y) + \bar{\imath}'\pi_n(Y') + d\pi_{n+1}(Y \times Y', Y \cup Y'), \quad n \geqslant 2,$$
where $\bar{\imath}$, $\bar{\imath}'$, d are isomorphisms. It should be noted that $\pi_n(Y \cup Y')$ is projected on to its summands $\pi_n(Y)$, $\pi_n(Y')$ by $\bar{\mu}, \bar{\mu}'$. It may also be noticed that $d\pi_2(Y \times Y', Y \cup Y')$ is the commutator subgroup $[\pi_1(Y), \pi_1(Y')]$ of $\pi_1(Y \cup Y')$.

7. The homotopy sequence of a triple. Let A be an arcwise-connected space and B, C closed arcwise-connected subspaces such that $C \subset B \subset A$. Then we have the sequence of homomorphisms
$$\ldots \rightarrow \pi_{n+1}(A, B) \xrightarrow{d^*_{n+1}} \pi_n(B, C) \xrightarrow{i^*_n} \pi_n(A, C) \xrightarrow{j^*_n} \pi_n(A, B) \xrightarrow{d^*_n} \ldots,$$
where i^*_n, j^*_n are injections induced by the identity maps
$$B, C \rightarrow A, C, \quad A, C \rightarrow A, B,$$
and d^*_{n+1} is the composition of $d_{n+1}: \pi_{n+1}(A, B) \rightarrow \pi_n(B)$, followed by the injection $\pi_n(B) \rightarrow \pi_n(B, C)$. This sequence is called *the homotopy sequence of the triple* (A, B, C). It reduces to the homotopy sequence of the pair (A, B) if C is the base-point for homotopy groups.

THEOREM 7·1. *The homotopy sequence of the triple* (A, B, C) *is exact.*

† The reader will by now have observed that arguments based on exact sequences frequently contain statements like 'i_n is on to $\pi_n(Y)$'. Following Bourbaki (and others) we will use the abbreviation 'i_n is on to' when it is clear what group is understood.

This follows from theorem 2·1, together with certain elementary facts. We will describe the proof that

$$i_n^{*-1}(0) = d_{n+1}^* \pi_{n+1}(A, B).$$

Consider the diagram

$$\begin{array}{ccccc}
 & & \pi_n(C) & = & \pi_n(C) \\
 & & \downarrow i'_n & & \downarrow i''_n \\
\pi_{n+1}(A,B) & \xrightarrow{d_{n+1}} & \pi_n(B) & \xrightarrow{i_n} & \pi_n(A) \\
 & & \downarrow j'_n & & \downarrow j''_n \\
 & & \pi_n(B,C) & \xrightarrow{i_n^*} & \pi_n(A,C) \\
 & & \downarrow d'_n & & \downarrow d''_n \\
 & & \pi_{n-1}(C) & = & \pi_{n-1}(C)
\end{array}$$

Here d_{n+1}, i_n are homomorphisms of the sequence of the pair (A, B); i'_n, j'_n, d'_n relate to the pair (B, C); and i''_n, j''_n, d''_n relate to the pair (A, C). We have, by definition, $d_{n+1}^* = j'_n d_{n+1}$, and the following 'commutativity' relations obviously hold:

$$i_n i'_n = i''_n, \quad i_n^* j'_n = j''_n i_n, \quad d''_n i_n^* = d'_n.$$

Now let $\alpha \in \pi_{n+1}(A, B)$. Then

$$i_n^* d_{n+1}^* \alpha = i_n^* j'_n d_{n+1} \alpha = j''_n i_n d_{n+1} \alpha = 0,$$

since $i_n d_{n+1} = 0$. Thus $d_{n+1}^* \pi_{n+1}(A, B) \subset i_n^{*-1}(0)$. Secondly, let $\alpha \in \pi_n(B, C)$, and let $i_n^* \alpha = 0$. Then $d'_n \alpha = d''_n i_n^* \alpha = 0$, so that, by exactness, $\alpha = j'_n \beta$, $\beta \in \pi_n(B)$. Then $0 = i_n^* \alpha = i_n^* j'_n \beta = j''_n i_n \beta$, so that, by exactness, $i_n \beta = i''_n \gamma$, $\gamma \in \pi_n(C)$. Thus

$$i_n(\beta - i'_n \gamma) = i_n \beta - i''_n \gamma = 0,$$

whence, by exactness, $\beta - i'_n \gamma = d_{n+1} \delta$, $\delta \in \pi_{n+1}(A, B)$, and $\alpha = j'_n \beta = j'_n(\beta - i'_n \gamma) = j'_n d_{n+1} \delta = d_{n+1}^* \delta$, since $j'_n i'_n = 0$. This shows that $i_n^{*-1}(0) \subset d_{n+1}^* \pi_{n+1}(A, B)$ and so completes the proof of the assertion that $i_n^{*-1}(0) = d_{n+1}^* \pi_{n+1}(A, B)$. The remaining assertions of the theorem may be proved similarly.

For the definition, properties and applications of the homotopy groups of a *triad*, see A. L. Blakers and W. S. Massey, 'The homotopy groups of a triad. I', *Ann. Math.* 53 (1951), 161–205.

CHAPTER V

FIBRE-SPACES

1. Definitions and fundamental theorems. Let $\phi: X \to Y$ be a map of X *on to* Y with the following properties:

(i) The counter-images $\phi^{-1}(y)$, $y \in Y$, are all homeomorphic to a certain space A.

(ii) There exists an open covering $\{U\}$ of Y, to each member of which corresponds a homeomorphism $\psi: U \times A \to \phi^{-1}(U)$, with $\psi(y, a) \in \phi^{-1}(y)$ if $y \in U$, $a \in A$.

Then we say† that X is a *fibre-space*, with *base-space* Y, *fibre* A, and *projection* (or *fibre-map*) $\phi: X \to Y$.

Examples. (i) $X = Y \times A$. Then $\phi(y, a) = y$, $y \in Y$, $a \in A$, any basis $\{U\}$ for the open sets of Y will do, and ψ is the identity.

(ii) X is a covering space of Y. Then A is a discrete set of points, ϕ is the covering map $X \to Y$, $\{U\}$ is a basis of neighbourhoods realizing the local homeomorphism between X and Y. The points of X lying over a given point of Y are 'indexed' by A, and if $(y, a) \in U(y_0) \times A$, then $\psi(y, a)$ is the image of y under the homeomorphism between $U(y_0)$ and a neighbourhood of the point of X lying over y_0 with index a.

(iii) X is a Möbius band, Y is a circle, A is a line segment (fig. 1).

(iv) X is a Klein bottle, Y is a circle, A is a circle (fig. 2).

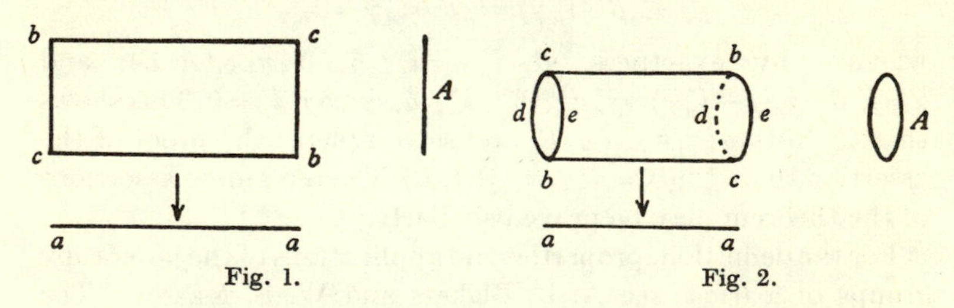

Fig. 1. Fig. 2.

† For the additional structure required to define a fibre-bundle, see Steenrod, *Topology of Fibre Bundles*, Princeton. We do not refer in this chapter to the *group* of the fibre.

(v) Compact Lie group, X, acting as a transitive group on a manifold, Y. The projection of X on Y is given by picking a particular $y_0 \in Y$, and defining $\phi(x)=x(y_0)$. The counter-images $\phi^{-1}(y)$ are then simply left cosets of the subgroup of X leaving y_0 fixed. The proof of the existence of the functions ψ in this case is given† in Chevalley, *Theory of Lie Groups*, no. 12, Princeton Mathematical Series, Prop. 1, p. 110.

It is interesting and important to study the homotopy relations of X, Y, and A. For example, a central question in fibre-theory is that of the existence (or non-existence) of a *cross-section*, that is, a map $f: Y \to X$ such that $\phi f: Y \to Y$ is the identity. We will see that light is thrown on this question by studying the homotopy relations of X, Y, A. More recently, a study has been made of the homology relations of X, Y, and A, but we do not discuss this here, except for an indication of its applicability in § 4.

The fundamental theorem on fibre-spaces may be formulated as an extension theorem.

THEOREM 1·1. *Let K be a (finite) simplicial complex and $L \subset K$ a subcomplex which is a deformation retract of K. Let $f_0: L \to X$ be a map such that $g_0=\phi f_0: L \to Y$ admits an extension to $g: K \to Y$. Then f_0 may be extended to $f: K \to X$ such that $g=\phi f$.*

Of course, any map of L into X (or Y) admits an extension to K. The force of the theorem is that we may 'cover' the given extension g by an extension f.

To prove the theorem, we first deal with the following special case. We take $K=\sigma \times I$, $L=\sigma \times 0 \cup \dot{\sigma} \times I$, where σ is a closed simplex,‡ and we assume that the unit interval may be divided at $0=t_0, t_1, \ldots, t_n=1$ in such a way that $g(\sigma \times \langle t_i, t_{i+1}\rangle)$ is contained in some open set, U_i, of the covering, for each $i=0, 1, \ldots, n-1$. Now suppose that f_0 has been extended to

$$f_i: \sigma \times \langle 0, t_i\rangle \cup \dot{\sigma} \times I \to X \quad \text{with} \quad \phi f_i = g \mid \sigma \times \langle 0, t_i\rangle \cup \dot{\sigma} \times I.$$

We put $\sigma_i=\sigma \times t_i$, $\tau_i=\dot{\sigma} \times \langle t_i, t_{i+1}\rangle$. Then

$$\phi f_i(\sigma_i \cup \tau_i)=g(\sigma_i \cup \tau_i) \subset g(\sigma \times \langle t_i, t_{i+1}\rangle) \subset U_i.$$

† Note that we may replace Y by its homeomorph, the factor-space X/X_0, where X_0 is the subgroup leaving y_0 fixed.

‡ If σ is 0-dimensional, $\dot{\sigma}$ is void, and $L=\sigma \times 0$.

Thus $f_i(\sigma_i \cup \tau_i) \subset \phi^{-1}(U_i)$. Now there exists a homeomorphism $\psi_i : U_i \times A \to \phi^{-1}(U_i)$. We define mappings $h_i : \sigma_i \cup \tau_i \to U_i$, $k_i : \sigma_i \cup \tau_i \to A$ by

$$f_i(p) = \psi_i(h_i(p), k_i(p)), \quad p \in \sigma_i \cup \tau_i.$$

Now $\phi\psi_i(h_i(p), k_i(p)) = h_i(p)$. Thus $h_i(p) = \phi f_i(p) = g(p)$, so that an extension of h_i to $\sigma \times \langle t_i, t_{i+1}\rangle$ is given by $h_i'(q) = g(q)$, $q \in \sigma \times \langle t_i, t_{i+1}\rangle$. Let θ be a retraction of $\sigma \times \langle t_i, t_{i+1}\rangle$ on to $\sigma_i \cup \tau_i$. Then an extension of k_i to $\sigma \times \langle t_i, t_{i+1}\rangle$ is given by $k_i'(q) = k_i\theta(q)$, $q \in \sigma \times \langle t_i, t_{i+1}\rangle$. We thus extend f_i to $f_{i+1} : \sigma \times \langle 0, t_{i+1}\rangle \cup \dot{\sigma} \times I \to X$ by defining $f_{i+1}(q) = \psi_i(g(q), k_i\theta(q))$, $q \in \sigma \times \langle t_i, t_{i+1}\rangle$. By construction, $\phi f_{i+1} = g \mid \sigma \times \langle 0, t_{i+1}\rangle \cup \dot{\sigma} \times I$. Repeating the argument as often as necessary, we finally construct the required map† f.

We now revert to the general case. We are given a map‡ $\rho : K \times I \to K$ with $\rho(p, 1) = p$, $\rho(q, t) = q$, $\rho(K \times 0) = L$, $p \in K$, $q \in L$, $t \in I$. Let

$$L^* = K \times 0 \cup L \times I, \quad K_r = K^r \cup L, \quad K_r^* = K \times 0 \cup K_r \times I$$

(so that $L^* = K_{-1}^*$); and define

$$f_0^* : L^* \to X, \quad g_0^* : L^* \to Y, \quad g^* : K \times I \to Y$$

by $f_0^* = f_0\rho$, $g_0^* = g_0\rho$, $g^* = g\rho$. We may further assume that K has been so finely subdivided that each of its simplexes σ has the property required in the proof of the special case. We may then assume that f_0^* has been extended to $f_{r+1}^* : K_r^* \to X$ with $\phi f_{r+1}^* = g^* \mid K_r^*$ (note that $g_0^* = \phi f_0^*$). If $r < \dim K$, we extend f_{r+1}^* (as in the special case) to each $\sigma^{r+1} \times I$ such that the interior of σ^{r+1} belongs to $K - L$, getting a map $f_{r+2}^* : K_{r+1}^* \to X$. We proceed in this way till we have a map $f^* : K \times I \to X$ with $\phi f^* = g^*$.

Finally, we define $f : K \to X$ by $f(p) = f^*(p, 1)$, $p \in K$. Then, if $q \in L$,

$$f(q) = f^*(q, 1) = f_0^*(q, 1) = f_0\rho(q, 1) = f_0(q),$$

and

$$\phi f(p) = \phi f^*(p, 1) = g^*(p, 1) = g\rho(p, 1) = g(p).$$

† The idea of the proof is simply to take advantage of the fact that, locally, X behaves like the product of Y and A.

‡ Note that, for convenience, we reverse the retracting homotopy $K \to K$.

Thus the map $f: K \to X$ satisfies the requirements of the theorem.

As a special case of this theorem we have

THEOREM 1·2. (Lifting homotopy theorem.) *Let $f_0: K \to X$ be a map of a (finite) simplicial complex K into X, and suppose that $g_0 = \phi f_0: K \to Y$ admits the homotopy $g_t: K \to Y$. Then f_0 admits the homotopy $f_t: K \to X$ such that $\phi f_t = g_t$.*

We simply replace K by $K \times I$, L by $K \times 0$ in theorem 1·1.

Let y_0 be chosen as base-point for homotopy groups† of Y, let A_0 be the fibre over y_0, and let $a_0 \in A_0$ be chosen as base-point for homotopy groups of X and X, A_0. Then ϕ induces

$$\phi_r: \pi_r(X, A_0, a_0) \to \pi_r(Y, y_0).$$

THEOREM 1·3. *ϕ_r is an isomorphism on to for all r.*

Of course, ϕ_1 is only asserted to be a (1-1) mapping of the set $\pi_1(X, A_0, a_0)$ on the set $\pi_1(Y, y_0)$.

First, let g be a map $E^r, S^{r-1} \to Y, y_0$, and let p_0 be the base-point in S^{r-1}. The map $f_0: p_0 \to X$ given by $f_0(p_0) = a_0$ is such that $\phi f_0: p_0 \to Y$ has the extension $g: E^r \to Y$. By theorem 1·1, f_0 has the extension $f: E^r \to X$ with $\phi f = g$. Since $gS^{r-1} = y_0$, $fS^{r-1} \subset A_0$, so that f is a map $E^r, S^{r-1}, p_0 \to X, A_0, a_0$, and ϕ_r is on to $\pi_r(Y, y_0)$.

Secondly, let $f': E^r, S^{r-1}, p_0 \to X, A_0, a_0$ be such that

$$\phi f' \sim 0: E^r, S^{r-1} \to Y, y_0.$$

Define $f_0: E^r \times 0 \cup p_0 \times I \to X$ by

$$f_0(p, 0) = f'(p), \quad p \in E^r, \quad f_0(p_0, t) = a_0, \quad t \in I.$$

The mapping $\phi f_0: E^r \times 0 \cup p_0 \times I \to Y$ admits the extension $g: E^r \times I, S^{r-1} \times I \to Y, y_0$, where g is the given homotopy of $\phi f'$ to the constant map. It is easy to prove (replacing E^r by I^r and p_0 by $(0, \ldots, 0)$) that $E^r \times 0 \cup p_0 \times I$ is a deformation retract of $E^r \times 1$. Thus we may extend f_0 to $f: E^r \times I \to X$ such that $\phi f = g$. Moreover, since $g(S^{r-1} \times I) = y_0$, $f(S^{r-1} \times I) \subset A_0$, so that f is a map

$$E^r \times I, S^{r-1} \times I, p_0 \times I \to X, A_0, a_0.$$

† We may insist, as heretofore, that Y be arcwise-connected (since otherwise we replace it by the arcwise-component of y_0), but we do not insist that X and A be arcwise-connected.

Now $g(E^r \times 1) = y_0$, so that $f(E^r \times 1) \subset A_0$. Thus the map $f'': E^r \to X$ given by $f''(p) = f(p, 1)$, $p \in E^r$, represents the 0 of $\pi_r(X, A_0, a_0)$. So, therefore, does $f': E^r, S^{r-1}, p_0 \to X, A_0, a_0$, whence ϕ_r is isomorphic and the theorem is proved.

COROLLARY 1·4. *If X is a fibre-space with fibre A_0, then $\pi_2(X, A_0)$ is Abelian.*

A particularly important special case of theorem 1·3 is the following:

THEOREM 1·5. *If $\tilde{Y}$ is a covering space of Y, and if $\tilde{y}_0$ is any point lying over y_0, then ϕ induces isomorphisms*

$$\phi_r^*: \pi_r(\tilde{Y}, \tilde{y}_0) \approx \pi_r(Y, y_0), \quad r \geqslant 2,$$

and ϕ_1^* is an isomorphism of $\pi_1(\tilde{Y}, \tilde{y}_0)$ into $\pi_1(Y, y_0)$.

For, in this case, the fibre A_0 is discrete. Thus, since S^{r-1} is connected if $r \geqslant 2$, any map $f: E^r, S^{r-1}, p_0 \to \tilde{Y}, A_0, \tilde{y}_0$ is really a map $f: E^r, S^{r-1} \to \tilde{Y}, \tilde{y}_0$, and any homotopy

$$f_t: E^r, S^{r-1}, p_0 \to \tilde{Y}, A_0, \tilde{y}_0$$

is really a homotopy $f_t: E^r, S^{r-1} \to \tilde{Y}, \tilde{y}_0$. Moreover, if

$$f_t: E^1, S^0, p_0 \to \tilde{Y}, A_0, \tilde{y}_0$$

is a homotopy of $f_0: E^1, S^0 \to \tilde{Y}, \tilde{y}_0$, then again f_t is really a homotopy of the form $f_t: E^1, S^0 \to \tilde{Y}, \tilde{y}_0$, since the path of any point under f_t is connected. Thus, if j_r is the injection

$$\pi_r(\tilde{Y}, \tilde{y}_0) \to \pi_r(\tilde{Y}, A_0, \tilde{y}_0),$$

j_r is an isomorphism (on to) if $r \geqslant 2$ and j_1 maps $\pi_1(\tilde{Y}, \tilde{y}_0)$ univalently† into $\pi_1(\tilde{Y}, A_0, \tilde{y}_0)$. Since $\phi_r^* = \phi_r j_r$, the theorem follows from theorem 1·3 (since ϕ_1^* is certainly a homomorphism).

COROLLARY 1·6. $\pi_r(S^1) = 0$, $r \geqslant 2$.

For the circle S^1 has the infinite line as universal covering space, and all the homotopy groups of the line are obviously trivial. The mapping ϕ from the line to the circle is simply $\phi(\alpha) = e^{i\alpha}$, α real.

COROLLARY 1·7. *If P^n is real projective n-space, then*

$$\pi_1(P^n) = Z_2, \quad \pi_r(P^n) \approx \pi_r(S^n), \quad r \geqslant 2.$$

† I.e. j_1 is (1-1); remember that $\pi_1(\tilde{Y}, A_0, \tilde{y}_0)$ has no group structure.

Theorem 1·3 may be relativized as follows:

THEOREM 1·8. *If Y_0 is a closed subset of Y containing y_0, and if $X_0 = \phi^{-1}(Y_0)$, then ϕ induces isomorphisms*

$$\phi'_r : \pi_r(X, X_0, a_0) \approx \pi_r(Y, Y_0, y_0).$$

If we specialize X to be a covering space, $\tilde{Y}$, of Y, and if we define $\tilde{Y}_0 = \phi^{-1}(Y_0)$, then we see that the covering map ϕ induces ismorphisms $\phi'_r : \pi_r(\tilde{Y}, \tilde{Y}_0, \tilde{y}_0) \approx \pi_r(Y, Y_0, y_0)$. If $r > 1$, we may replace $\tilde{Y}_0$ by the arcwise component containing $\tilde{y}_0$.

2. The Hopf fibrings. We will define a map $\phi : S^3 \to S^2$ and prove that it is a projection of the fibre-space S^3 on to the base-space S^2 with fibres S^1. We represent S^3 as the space of two complex variables (z_1, z_2) subject to $z_1\bar{z}_1 + z_2\bar{z}_2 = 1$, and we represent S^2 as the complex projective line whose points are classes of pairs of complex numbers, $[z_1, z_2]$, not both zero, where we put $[z_1, z_2]$ and $[z'_1, z'_2]$ into the same class if there exists a complex number λ such that $z'_1 = \lambda z_1$, $z'_2 = \lambda z_2$. We then define $\phi : S^3 \to S^2$ by $\phi(z_1, z_2) = [z_1, z_2]$. Since we may 'normalize' $[z_1, z_2]$ by dividing by $(z_1\bar{z}_1 + z_2\bar{z}_2)^{\frac{1}{2}}$, it follows that ϕ maps S^3 on to S^2. Also (z_1, z_2) and (z'_1, z'_2) have the same image under ϕ if and only if there exists λ such that $z'_1 = \lambda z_1$, $z'_2 = \lambda z_2$; but then

$$1 = z'_1\bar{z}'_1 + z'_2\bar{z}'_2 = \lambda\bar{\lambda}(z_1\bar{z}_1 + z_2\bar{z}_2) = \lambda\bar{\lambda},$$

so that $|\lambda| = 1$. Thus if $\phi(z_1, z_2) = [z_1, z_2]$, then $\phi^{-1}[z_1, z_2]$ consists of those points which are obtained from (z_1, z_2) by multiplying each coordinate by $e^{i\theta}$, $-\pi < \theta \leqslant \pi$. In other words, the counter-images of points of S^2 are great circles on S^3. We have an open covering of S^2 consisting of $S^2 - [0, 1]$ and $S^2 - [1, 0]$. If we write μ for z_1/z_2, putting $\mu = \infty$ if $z_2 = 0$, then we may call these open sets $S^2 - 0$ and R^2. We define the homeomorphism

$$\psi_1 : (S^2 - 0) \times S^1 \to \phi^{-1}(S^2 - 0)$$

by $$\psi_1(\mu, \theta) = \left(\frac{e^{i\theta}}{\sqrt{(1 + 1/|\mu|^2)}}, \frac{e^{i\theta}}{\mu\sqrt{(1 + 1/|\mu|^2)}}\right), \quad \mu \neq \infty$$

$$\psi_1(\infty, \theta) = (e^{i\theta}, 0)$$

and we define the homeomorphism $\psi_2 : R^2 \times S^1 \to \phi^{-1}(R^2)$ by

$$\psi_2(\mu, \theta) = \left(\frac{\mu e^{i\theta}}{\sqrt{(1 + |\mu|^2)}}, \frac{e^{i\theta}}{\sqrt{(1 + |\mu|^2)}}\right).$$

Thus the map ϕ has the required properties, so that it induces isomorphisms $\phi_r:\pi_r(S^3,S^1)\approx\pi_r(S^2)$. Since S^1 is contractible over S^3 to a point, we have†

$$\pi_r(S^3,S^1)\approx\pi_r(S^3)+\pi_{r-1}(S^1).$$

Thus, if $r\geqslant 3$, it follows from corollary 1·6 that

$$\pi_r(S^3)\approx\pi_r(S^3,S^1).$$

Moreover, the isomorphism is induced by the injection

$$j_r:\pi_r(S^3)\to\pi_r(S^3,S^1).$$

We conclude then

THEOREM 2·1. *The mapping* $\phi:S^3\to S^2$ *induces isomorphisms* $\phi_r^*:\pi_r(S^3)\approx\pi_r(S^2)$, $r\geqslant 3$.

COROLLARY 2·2. $\pi_3(S^2)$ *is cyclic infinite, generated by the class of the 'Hopf' map* $\phi:S^3\to S^2$.

In the same way, we may define a projection ϕ of S^7 on S^4 with fibre S^3. We represent S^7 as the space of two quaternionic variables (q_1,q_2) such that $|q_1|^2+|q_2|^2=1$ and S^4 as the 'quaternionic projective line' whose points are classes of pairs $[q_1,q_2]$, not both of q_1,q_2 being zero, $[q_1,q_2]$ and $[q_1',q_2']$ going into the same class if there exists a quaternion q such that $q_1'=qq_1$, $q_2'=qq_2$. Just as before, the map $(q_1,q_2)\to[q_1,q_2]$ maps S^7 on S^4 and the counter-images are obtained from a single point in the counter-image by multiplying its coordinates (on the left) by quaternions of unit norms. Thus the counter-images are all great 3-spheres. Again we take $S^4-[0,1]$, $S^4-[1,0]$ as the open covering of S^4 and the homeomorphisms ψ_1,ψ_2 are defined analogously.

Since S^3 is contractible over S^7 to a point, we have

$$\pi_r(S^7,S^3)\approx\pi_r(S^7)+\pi_{r-1}(S^3).$$

This, together with the isomorphism $\phi_r:\pi_r(S^7,S^3)\approx\pi_r(S^4)$, gives

THEOREM 2·3. $\pi_r(S^4)\approx\pi_r(S^7)+\pi_{r-1}(S^3)$.

COROLLARY 2·4. $\pi_5(S^4)\approx\pi_4(S^3)$, $\pi_6(S^4)\approx\pi_5(S^3)$.

COROLLARY 2·5. $\pi_7(S^4)$ *has a cyclic infinite direct summand generated by the class of the 'Hopf' map* $S^7\to S^4$.

† See Chapter IV, theorem 5·1.

The constructions given above may be generalized as follows (we carry out the generalization explicitly in the complex case†).

We may represent any point of S^{2n-1} by the n-tuple of complex numbers $(z_1, \ldots, z_n)$ with $|z_1|^2 + \ldots + |z_n|^2 = 1$. Complex projective space of $(n-1)$ complex dimensions, M_n, $n \geqslant 2$, may be represented as the space of classes of n-tuples $[z_1, \ldots, z_n]$, not all the z_i being zero, where we put $[z_1, \ldots, z_n]$ and $[z'_1, \ldots, z'_n]$ in the same class if there exists a complex number λ such that $z'_i = \lambda z_i$, $i = 1, \ldots, n$. Then the map $(z_1, \ldots, z_n) \rightarrow [z_1, \ldots, z_n]$ is a projection of S^{2n-1} on complex projective space M_n, and the fibres are great circles on S^{2n-1}. Thus, just as in the special case $n = 2$, we get the isomorphism $\pi_r(M_n) \approx \pi_r(S^{2n-1}) + \pi_{r-1}(S^1)$, whence

THEOREM 2·6. *$\pi_r(M_n) \approx \pi_r(S^{2n-1})$, $r \geqslant 3$, and $\pi_2(M_n) = Z_\infty$.*

COROLLARY 2·7. *If M_3 is the complex projective plane, then*

$$\pi_3(M_3) = \pi_4(M_3) = 0, \quad \pi_2(M_3) = \pi_5(M_3) = Z_\infty.$$

In the same way, we may define a projection of S^{4n-1} on to quaternionic projective space of $(n-1)$ quaternionic dimensions in which the fibres are great 3-spheres on S^{4n-1}. Again, just as in the special case $n = 2$, we get the isomorphism

$$\pi_r(Q_n) \approx \pi_r(S^{4n-1}) + \pi_{r-1}(S^3),$$

where Q_n is quaternionic projective space of $(n-1)$ quaternionic dimensions (and therefore of $(4n-4)$ topological dimensions).

THEOREM 2·8. *$\pi_r(Q_n) \approx \pi_r(S^{4n-1}) + \pi_{r-1}(S^3)$.*

COROLLARY 2·9. *$\pi_r(Q_n) \approx \pi_{r-1}(S^3)$, $r \leqslant 4n-2$, and $\pi_{4n-1}(Q_n)$ contains a cyclic infinite direct summand generated by the class of the 'Hopf' map $S^{4n-1} \rightarrow Q_n$.*

The Cayley numbers form a non-associative division algebra of order 8. If we attempt to carry out a precisely similar procedure as with the complex numbers and the quaternions to obtain a projection of S^{15} on S^8 we meet a snag in attempting to define 'Cayley projective space', due to the non-associativity of the multiplication of Cayley numbers. However, we can modify

† The 'generalization in the real case' gives the double covering $S^{n-1} \rightarrow P^{n-1}$, where P^{n-1} is real projective $(n-1)$-space. See corollary 1·7.

the definitions slightly to get a representation of S^{15} as a fibre-space over S^8 with fibres great 7-spheres. However, this modification does not enable us to generalize the procedure as we did in the arguments leading to theorems 2·6 and 2·8.

We take S^{15} as the space of pairs of Cayley numbers (X, Y) with $|X|^2+|Y|^2=1$. The 'Cayley lines' $Y=CX$, together with the 'line' $X=0$, are mutually disjoint (except, of course, for the 'origin'), and span R^{16}. We associate with all points of S^{15} on $Y=CX$ the Cayley number C and with all points on $X=0$ the point at ∞ (which closes Euclidean 8-space R^8 to S^8). In this way we obtain a mapping of S^{15} on S^8 and the counter-images are 'Cayley lines'. The equation $Y=CX$, for fixed C, is actually a set of eight homogeneous linear equations in the 16 coordinates of R^{16}, and thus the 'Cayley line' $Y=CX$ is an 8-dimensional hyperplane in R^{16} through the centre of S^{15} and thus intersects S^{15} in a great 7-sphere. As before, we take R^8 and† S^8-0 as our covering of S^8 and define

$$\left.\begin{array}{l}
\psi_1: R^8\times S^7\to\phi^{-1}(R^8) \quad \text{by} \\
\psi_1(C,D)=\left(\dfrac{D}{\sqrt{(1+|C|^2)}}, \dfrac{CD}{\sqrt{(1+|C|^2)}}\right) \\
\text{and} \quad \psi_2:(S^8-0)\times S^7\to\phi^{-1}(S^8-0) \quad \text{by} \\
\psi_2(C,D)=\left(\dfrac{\dfrac{D}{C}}{\sqrt{\left(1+\dfrac{1}{|C|^2}\right)}}, \dfrac{D}{\sqrt{\left(1+\dfrac{1}{|C|^2}\right)}}\right) \\
\psi_2(\infty,D)=(0,D),
\end{array}\right\} \begin{array}{l}(C \text{ a Cayley} \\ \text{number, } D \\ \text{a Cayley} \\ \text{number of} \\ \text{unit norm})\end{array}$$

where D/C means the solution A of the equation $CA=D$.

Thus we get a 'Hopf' map $S^{15}\to S^8$ with fibres S^7, so that

THEOREM 2·10. $\pi_r(S^8)\approx\pi_r(S^{15})+\pi_{r-1}(S^7)$.

COROLLARY 2·11. $\pi_r(S^8)\approx\pi_{r-1}(S^7)$, $r\leqslant 14$.

COROLLARY 2·12. $\pi_{15}(S^8)$ *contains a cyclic infinite direct summand generated by the class of the 'Hopf' map* $S^{15}\to S^8$.

† 0 is the origin of coordinates in R^8. When R^n is closed by a point at infinity it is written by some authors as Z^n.

All the fibrings dealt with in this section were described, essentially, by Hopf in his paper 'Über die Abbildungen von Sphären auf Sphären niedrigeres Dimension', *Fundament. Math.* (1935), pp. 427–40, though he did not there draw the full information available on the homotopy groups of spheres and projective spaces.

3. Fibre-spaces over spheres. The Hopf fibre-maps $S^3 \to S^2$, $S^7 \to S^4$, $S^{15} \to S^8$ are examples of a fibre-space over a sphere. In this section we consider simplifications which may be made in the general theory if the base-space is a sphere. Before we do this, however, we introduce the *homotopy sequence of a fibre-space.*

Let $\phi: X \to Y$ be a projection of X on to Y with fibre A. Then if A_0 is the fibre over $y_0 \in Y$, we know that ϕ induces isomorphisms $\phi_r: \pi_r(X, A_0) \approx \pi_r(Y)$. Given the exact sequence of the pair (X, A_0), namely,

$$\ldots \longrightarrow \pi_{n+1}(X, A_0) \xrightarrow{d_{n+1}} \pi_n(A_0) \xrightarrow{i_n} \pi_n(X) \xrightarrow{j_n} \pi_n(X, A_0) \to \ldots,$$

we derive a new exact sequence by writing

$$\bar{j}_n = \phi_n j_n, \quad \bar{d}_{n+1} = d_{n+1} \phi_{n+1}^{-1},$$

namely,

$$\ldots \longrightarrow \pi_{n+1}(Y) \xrightarrow{\bar{d}_{n+1}} \pi_n(A_0) \xrightarrow{i_n} \pi_n(X) \xrightarrow{\bar{j}_n} \pi_n(Y) \to \ldots, \quad (3{\cdot}1)$$

called *the homotopy sequence of the fibre-space X.*† It is trivial to show that it is exact.

Now let us suppose that $Y = S^m$. Since $\pi_r(S^m) = 0$, $r < m$, it follows from the exactness of the sequence (3·1) that

$$\pi_n(A_0) \approx \pi_n(X), \quad n \leqslant m-2. \quad (3{\cdot}2)$$

It also follows that i_{m-1} maps $\pi_{m-1}(A_0)$ on to $\pi_{m-1}(X)$ and that the kernel of i_{m-1} is $\bar{d}_m \pi_m(S^m)$. Since $\pi_m(S^m)$ is cyclic infinite, $\bar{d}_m \pi_m(S^m)$ is cyclic. Let it be generated by η, so that $\eta \in \pi_{m-1}(A_0)$. η is called by some authors the *characteristic class of the fibre-space X over S^m.* To find a representative of η, we choose a map

$$f: E^m, S^{m-1} \to X, A_0$$

representing $\phi_m^{-1}(1)$, where 1 generates $\pi_m(S^m)$, and then cut f down to S^{m-1}. The map f may be taken to be homeomorphic on the interior of E^m.

† The projection $\phi: X \to Y$ induces the homomorphism $\bar{j}_n$; $\bar{j}_1$ is, of course, a homomorphisn, although $\pi_1(X, A_0)$ has not been given a group structure.

THEOREM 3·3. *The fibre-space X admits a cross-section if and only if $\eta=0$.*

We recall that a cross-section is a map $g: Y\to X$ such that $\phi g=1$. Suppose that a cross-section exists. Now the map ϕ induces the homomorphisms $\bar{\jmath}_n:\pi_n(X)\to\pi_n(S^m)$. Since ϕ has a right-inverse, these homomorphisms must be on to $\pi_n(S^m)$. In particular, $\bar{\jmath}_m$ maps $\pi_m(X)$ on to $\pi_m(S^m)$, so that, by exactness, $\bar{d}_m$ maps $\pi_m(S^m)$ on to 0, or $\eta=0$.

Now let $\eta=0$. Then, by exactness, $\bar{\jmath}_m$ maps $\pi_m(X)$ on to $\pi_m(S^m)$. This implies that there exists a map $g': S^m\to X$ such that $\phi g'\sim 1: S^m\to S^m$; but then theorem 1·2 ensures that $g'\sim g: S^m\to X$ with $\phi g=1$, and g is the required cross-section.

THEOREM 3·4. *If the projection $X\to Y$ admits a cross-section, then*

$$\pi_n(X)\approx\pi_n(Y)+\pi_n(A_0),\quad n\geqslant 2.$$

For if $g: Y\to X$ is a map such that $\phi g=1$, and if g induces $\gamma_n:\pi_n(Y)\to\pi_n(X)$, then $\bar{\jmath}_n\gamma_n=1:\pi_n(Y)\approx\pi_n(Y)$, whence, if $n\geqslant 2$, $\pi_n(X)=\gamma_n\pi_n(Y)+\bar{\jmath}_n^{-1}(0)$, and γ_n is isomorphic. Now $\bar{\jmath}_n^{-1}(0)=i_n\pi_n(A_0)$, and it follows, by exactness, from the fact that $\bar{\jmath}_{n+1}$ is on to $\pi_{n+1}(Y)$, that i_n is isomorphic. This proves the theorem.†

We study as an example the space of unit tangent vectors on S^n. The existence of a cross-section is then equivalent to the existence of a field of tangent vectors on S^n. In fact, we generalize the problem by considering *Stiefel manifolds*. The Stiefel manifold $V_{n,m}$, $n>m$, is the manifold whose points are systems of m unit orthogonal vectors in R^n. To $a\in V_{n,m}$ corresponds $\phi^{(k)}(a)\in V_{n,k}$, given by the first k vectors of a. Given $\phi^{(k)}(a)$, we are free to choose the remaining $(m-k)$ vectors of a from the $(n-k)$-dimensional subspace orthogonal to $\phi^{(k)}(a)$. Thus it may be shown that $\phi^{(k)}$ is a projection of $V_{n,m}$ on $V_{n,k}$ with fibre $V_{n-k,m-k}$.

If we fix an orthogonal coordinate system in R^n, then $a\in V_{n,m}$ is given by a matrix A of m rows and n columns satisfying $AA'=I_m$, where I_m is the m-rowed unit matrix. Notice that

† If we take $X=S^3$, $Y=S^2$, $A_0=S^1$, $n=2$, we see that the projection $\phi: S^3\to S^2$ does not admit a cross-section. Neither, in fact, do any of the Hopf fibrings.

$V_{n-k,m-k}$ becomes a subspace of $V_{n,m}$ when the matrix $B \in V_{n-k,m-k}$ is identified with

$$\begin{Vmatrix} I_k & 0_{k,n-k} \\ 0_{m-k,k} & B \end{Vmatrix} \in V_{n,m}.$$

If $m=1$, then the points of $V_{n,1}$ are units vectors in R^n, and $V_{n,1}$ is thus homeomorphic to S^{n-1}. Thus we may identify $V_{n+1,1}$ with S^n, and $V_{n+1,2}$ with the space of unit tangent vectors to S^n; for we simply translate the second vector to the end-point of the first. Similarly, $V_{n+1,m+1}$ is the space of sets of m unit orthogonal tangent vectors to S^n. Moreover, the map $\phi : V_{n+1,m+1} \to S^n$, which simply associates with a set of tangent vectors the point at which they act, is the same as the projection $\phi^{(1)} : V_{n+1,m+1} \to V_{n+1,1}$, for each consists in picking out the first row of the matrix A.

Now let $a \in V_{n,n-1}$ be represented by the matrix A. Then A may be completed in precisely one way by the addition of an nth row to give an orthogonal matrix of determinant $+1$. In this way $V_{n,n-1}$ may be identified with the unimodular orthogonal group,† or *rotation group*, Ω_n. The projection $\phi : \Omega_{n+1} \to S^n$, with fibre Ω_n, may then be regarded as a special case of example (v), at the beginning of the chapter, where we pick out $p_0 = (1, 0, \ldots, 0) \in S^n$ and define $\phi(\omega) = \omega(p_0)$, $\omega = \Omega_{n+1}$.

If we consider the projection $\phi : V_{n+1,m+1} \to S^n$, with fibre $V_{n,m}$, we see that

$$\pi_r(V_{n,m}) \approx \pi_r(V_{n+1,m+1}), \quad r < n-1, \tag{3·5}$$

and

$$\pi_{n-1}(V_{n,m})/\bar{d}_n \pi_n(S^n) \approx \pi_{n-1}(V_{n+1,m+1}). \tag{3·6}$$

By theorem 3·3, the question of the existence, or otherwise, of a field of m unit orthogonal tangent vectors on S^n is just the question of whether the generating element of $\bar{d}_n \pi_n(S^n)$ is or is not zero.

Now let E^n be the set of points $x = (x_0, \ldots, x_n)$, $x_0^2 + \ldots + x_n^2 = 1$, $x_0 \geqslant 0$, and let S^{n-1} be its boundary, given by $x \in E^n$, $x_0 = 0$.‡ Let $u : E^n \to V_{n+1,m+1}$ be the map given by the matrix

$$u_{ij}(x) = \delta_{ij} - 2x_i x_j, \quad i = 0, \ldots, m, \; j = 0, \ldots, n,$$

† The rotation group is the identity component of the full orthogonal group. Thus, if we extend the definition of homotopy group in the obvious way, the homotopy groups of the rotation group and the orthogonal group are the same.

‡ Note the change of 'co-ordinate system'.

where δ_{ij} is the Kronecker symbol. Then it is easily verified that $\|u_{ij}(x)\| \in V_{n+1,m+1}$ and that ϕu maps $E^n - S^{n-1}$ topologically on to $S^n - (1,0,\ldots,0)$ and $\phi u(S^{n-1}) = (1,0,\ldots,0)$. (Note that we obtain ϕ by restricting attention to $i=0$.) Thus if we restrict u to S^{n-1} we get a representative of a generator of $\bar{d}_n \pi_n(S^n)$. It is the map $t: S^{n-1} \to V_{n,m}$ given by

$$t_{ij}(x) = \delta_{ij} - 2x_i x_j, \quad i=1,\ldots,m,\ j=1,\ldots,n,$$

where $x = (x_1, \ldots, x_n) \in S^{n-1}$, so that $x_1^2 + \ldots + x_n^2 = 1$.

Let us put $m=1$. Then t is a mapping of S^{n-1} into S^{n-1} given by

$$t(x_1, \ldots, x_n) = (1 - 2x_1^2, -2x_1x_2, \ldots, -2x_1x_n).$$

Now† $t \mid E_+^{n-1}$ $(x_1 \geqslant 0)$ is a map of E_+^{n-1} on to S^{n-1} which is a homeomorphism of the interior of E_+^{n-1} on to $S^{n-1} - (1,0,\ldots,0)$ and t maps the bounding S^{n-2} to $(1,0,\ldots,0)$. Also $(x_1,\ldots,x_n)$ and $(x_1',\ldots,x_n')$ have the same image under t if and only if $x_i' = -x_i$, $i=1,\ldots,n$. Since the map $(x_1,\ldots,x_n) \to (-x_1,\ldots,-x_n)$ has degree $(-1)^n$, it follows that t has degree $1+(-1)^n$. Thus we have proved

THEOREM 3·7. *S^n admits a field of tangent vectors if and only if n is odd.*

For a map $t: S^{n-1} \to S^{n-1}$ is nullhomotopic if and only if it has degree 0. Further results along these lines have been obtained by G. W. Whitehead, B. Eckmann, N. E. Steenrod and others. The reader is referred to N. E. Steenrod, *Topology of Fibre Bundles*, Princeton, p. 142.

We note that we may conclude from theorems 3·4 and 3·7 that $\pi_r(V_{n+1,2}) \approx \pi_r(S^n) + \pi_r(S^{n-1})$, $r \geqslant 2$, n odd. It is clear from (3·5) that $\pi_1(V_{n+1,2}) = 0$, since $n \geqslant 3$, so that, *if n is odd*,

$$\pi_r(V_{n+1,2}) \approx \pi_r(S^n) + \pi_r(S^{n-1}). \tag{3·8}$$

We return later to the consideration of $\pi_r(V_{n+1,2})$ when n is even.

THEOREM 3·9. *If n is even, the injection*

$$i_n : \pi_n(V_{n,m}) \to \pi_n(V_{n+1,m+1})$$

is on to $\pi_n(V_{n+1,m+1})$.

† We use the symbol E_+^{n-1} here for the hemisphere of S^{n-1} given by $x_1 \geqslant 0$.

It follows from the exact sequence,

$$\pi_n(V_{n,m}) \xrightarrow{i_n} \pi_n(V_{n+1,m+1}) \xrightarrow{\bar{j}_n} \pi_n(S^n) \xrightarrow{\bar{d}_n} \pi_{n-1}(V_{n,m}),$$

that we have to prove that $\bar{d}_n$ is an isomorphism. Now the projection $V_{n,m} \to S^{n-1}$ transforms the generator of $\bar{d}_n \pi_n(S^n)$ into the class, $\epsilon \pi_{n-1}(S^{n-1})$, represented by the map $t: S^{n-1} \to S^{n-1}$ discussed in the proof of theorem 3·7. We saw that, if n is even, t represents the element $2 \epsilon \pi_{n-1}(S^{n-1})$. Thus the image of $\bar{d}_n \pi_n(S^n)$ in the homomorphism $\pi_{n-1}(V_{n,m}) \to \pi_{n-1}(S^{n-1})$ induced by the projection $V_{n,m} \to S^{n-1}$ is a cyclic infinite group, so that $\bar{d}_n \pi_n(S^n)$ is cyclic infinite, implying that $\bar{d}_n$ is an isomorphism. This proves the theorem.

So far we have been considering the projection $\phi: V_{n+1,m+1} \to S^n$ with fibre $V_{n,m}$. It is also instructive to consider the projection $\phi' = \phi^{(m-1)}: V_{n,m} \to V_{n,m-1}$ with fibre $V_{n-m+1,1} = S^{n-m}$. From the 'fibre'-sequence

$$\pi_r(S^{n-m}) \xrightarrow{i_r} \pi_r(V_{n,m}) \xrightarrow{\bar{j}_r} \pi_r(V_{n,m-1}) \xrightarrow{\bar{d}_r} \pi_{r-1}(S^{n-m}),$$

it follows that

$$\pi_r(V_{n,m}) \approx \pi_r(V_{n,m-1}), \quad r < n-m, \tag{3·10}$$

and

$$\pi_{n-m}(V_{n,m})/i_{n-m}\pi_{n-m}(S^{n-m}) \approx \pi_{n-m}(V_{n,m-1}). \tag{3·11}$$

Suppose now we put $m = n-1$. Then $S^{n-m} = S^1$ and $V_{n,n-1}$ is the rotation group Ω_n. Since $\pi_r(S^1) = 0$, $r \geqslant 2$, we have

$$\pi_r(\Omega_n) \approx \pi_r(V_{n,n-2}), \quad r \geqslant 3. \tag{3·12}$$

Ω_2 is simply the circle; it may also be shown that Ω_3 is projective 3-space. (For example, any rotation of S^2 is a rotation about a line; if we mark off the segment $\langle -\pi, \pi \rangle$ on this line, a point on the segment uniquely determines a rotation provided we identify $-\pi$ with π.) Thus $\pi_1(\Omega_2) = Z_\infty$, $\pi_1(\Omega_3) = Z_2$, and, by (3·5),

$$\pi_1(\Omega_n) = Z_2, \quad n \geqslant 3. \tag{3·13}$$

Again, $\pi_2(\Omega_2) = 0$, $\pi_2(\Omega_3) = 0$, $\pi_2(\Omega_4) = 0$ by (3·6), and, by (3·5),

$$\pi_2(\Omega_n) = 0, \quad \text{all } n. \tag{3·14}$$

Now let us represent S^3 as the space of quaternions of unit norm. Then $\phi: \Omega_4 \to S^3$ is defined by $\phi(r) = r(1)$, $r \in \Omega_4$. We may define a map $f: S^3 \to \Omega_4$ by $f(q)(q') = qq'$, $q, q' \in S^3$. Since $|qq'| = 1$, it is clear that $f(q) \in \Omega_4$. Also $\phi f(q) = f(q)(1) = q$, so that f is a cross-section. Thus† $\pi_n(\Omega_4) \approx \pi_n(\Omega_3) + \pi_n(S^3)$, $n \geqslant 2$, or

$$\pi_n(\Omega_4) \approx \pi_n(S^3) + \pi_n(S^3), \quad n \geqslant 2. \tag{3·15}$$

Thus $\pi_3(\Omega_2) = 0$, $\pi_3(\Omega_3) = Z_\infty$, $\pi_3(\Omega_4) = Z_\infty + Z_\infty$ and, by (3·6), $\pi_3(\Omega_5)$ is a homomorphic image of $Z_\infty + Z_\infty$ whose kernel is Z_∞ or 0. By (3·5), all the Ω_n, $n \geqslant 5$, have isomorphic third homotopy groups.

The map $f: S^3 \to \Omega_4$, given above, clearly represents a generator, σ, of $\pi_3(\Omega_4)$. The other generator may be represented by a map $S^3 \to \Omega_3$. In fact, let $g: S^3 \to \Omega_4$ be given by $g(q)(q') = qq'q^{-1}$, $q, q' \in S^3$. Then if S^2 is represented as the set of 'pure imaginary' quaternions (i.e. quaternions of the form $0\mathbf{e} + a\mathbf{i} + b\mathbf{j} + c\mathbf{k}$, where $\mathbf{e}$, $\mathbf{i}$, $\mathbf{j}$, $\mathbf{k}$ are the quaternionic units) of unit norm, it follows from the fact that $g(q)$ is a rotation and $q1q^{-1} = 1$ that $g(q)$ actually belongs to Ω_3. Moreover, $g(q)$ is the identity if and only if $q = 1$ or -1. Thus g is just the double covering‡ of S^3 on to projective 3-space, so that g represents a generator of $\pi_3(\Omega_3)$. Thus if g represents $\rho \in \pi_3(\Omega_4)$, we have shown that $\pi_3(\Omega_4) = Z_\infty + Z_\infty$, generated by ρ and σ.

THEOREM 3·16. *$\pi_3(\Omega_5)$ is cyclic infinite, generated by the image of σ. Moreover, the kernel of the injection $\pi_3(\Omega_4) \to \pi_3(\Omega_5)$ is generated by $-\rho + 2\sigma$.*

The proof depends on the following lemma, whose interest extends considerably beyond our particular application:

LEMMA 3·17. *If G is a topological group with identity element e, and if $f_i: I^n, \dot{I}^n \to G, e$ represent $\alpha_i \in \pi_n(G, e)$, $i = 1, 2$, then $\alpha_1 + \alpha_2$ is represented by the map $f_1 f_2: I^n, \dot{I}^n \to G, e$, given by*

$$f_1 f_2(t) = f_1(t) f_2(t), \quad t \in I^n.$$

† Note that this implies that the injection $\pi_n(\Omega_3) \to \pi_n(\Omega_4)$ maps $\pi_n(\Omega_3)$ isomorphically.

‡ It is not difficult to show that g maps S^3 on to Ω_3.

Let $f_1' = f_1 + 0, f_2' = 0 + f_2$, i.e. $f_1'(t_1, \ldots, t_n) = f_1(2t_1, \ldots, t_n), 0 \leqslant t_1 \leqslant \frac{1}{2}$, $f_1'(t_1, \ldots, t_n) = e, \frac{1}{2} \leqslant t_1 \leqslant 1$, and similarly for f_2'. Then

$$f_i' \sim f_i : I^n, \dot{I}^n \to G, e,$$

whence $f_1' f_2' \sim f_1 f_2 : I^n, \dot{I}^n \to G, e$. But $f_1' f_2' = f_1 + f_2$. Thus

$$f_1 + f_2 \sim f_1 f_2 : I^n, \dot{I}^n \to G, e,$$

whence $f_1 f_2$ represents $\alpha_1 + \alpha_2$.

Reverting to the theorem, we observe that the map

$$h(q) = g(q)^{-1} (f(q))^2 \in \Omega_4$$

is given by $(h(q))(q') = g(q)^{-1}(q^2 q') = qq'q$. On the other hand, by the lemma $h : S^3 \to \Omega_4$ represents $-\rho + 2\sigma \in \pi_3(\Omega_4)$.

Now the kernel of the injection $\pi_3(\Omega_4) \to \pi_3(\Omega_5)$ is the cyclic group $\bar{d}_4 \pi_4(S^4)$. We have seen that, if Ω_4 is identified with $V_{4,3}$, then a generator of $\bar{d}_4 \pi_4(S^4)$ is represented by the map

$$t : S^3 \to V_{4,3}$$

given by

$$t_{ij}(x) = \delta_{ij} - 2x_i x_j, \quad i = 1, 2, 3; j = 1, 2, 3, 4,$$

where $x = (x_1, x_2, x_3, x_4) \in S^3$.

If we identify $V_{4,3}$ with Ω_4, as described above, by adding a fourth row to the matrix of $V_{4,3}$, we get

$$t_{4j}(x) = 2x_4 x_j - \delta_{4j}, \quad j = 1, 2, 3, 4.$$

Thus

$$t(x) = \begin{Vmatrix} 1 & & & \\ & 1 & & \\ & & 1 & \\ & & & -1 \end{Vmatrix} \| \delta_{ij} - 2x_i x_j \|, \quad i, j = 1, 2, 3, 4; \ t : S^3 \to \Omega_4.$$

If we replace t by t', given by

$$t'(x) = \begin{Vmatrix} -1 & & & \\ & 1 & & \\ & & 1 & \\ & & & 1 \end{Vmatrix} \| \delta_{ij} - 2x_i x_j \|,$$

then it is clear that t' also represents a generator of $\bar{d}_4 \pi_4(S^4)$ and further has the property that $t'(1, 0, 0, 0)$ is the identity matrix. We will show that the maps h and t' coincide. First we identify the quaternion q, $=x_1+x_2\mathbf{i}+x_3\mathbf{j}+x_4\mathbf{k}$, with $|q|=1$, with the point $(x_1,x_2,x_3,x_4)\in S^3$. Then we observe that since $V_{4,3}$ is projected on to S^3 by taking the first row of the matrix, and since Ω_4 is projected on to S^3 by taking $\phi(r)=r(1)$, where $r\in\Omega_4$, and 1 is the unit quaternion, $=(1,0,0,0)$, we must for consistency have Ω_4 operating on the right on row vectors corresponding to points of S^3. Thus we must verify that, if $q=(x_1,x_2,x_3,x_4)$, $q'=(x_1',x_2',x_3',x_4')$, then

$$qq'q=(x_1',x_2',x_3',x_4')\left\|\begin{matrix}-1&&&\\&1&&\\&&1&\\&&&1\end{matrix}\right\|\,\|\delta_{ij}-2x_ix_j\|$$

$$=(-x_1',x_2',x_3',x_4')\,\|\delta_{ij}-2x_ix_j\|.$$

It follows by linearity that it is sufficient to verify this if $q'=1,\mathbf{i},\mathbf{j},\mathbf{k}$, and by symmetry that it is sufficient to verify it if $q'=1,\mathbf{i}$. We recall that $|q|=1$.

Now
$$q^2=x_1^2-x_2^2-x_3^2-x_4^2+2x_1x_2\mathbf{i}+2x_1x_3\mathbf{j}+2x_1x_4\mathbf{k}$$
$$=(2x_1^2-1,2x_1x_2,2x_1x_3,2x_1x_4),$$
$$=(-1,0,0,0)\,\|\delta_{ij}-2x_ix_j\|.$$

Again,

$$qiq=(-x_2+x_1\mathbf{i}+x_4\mathbf{j}-x_3\mathbf{k})(x_1+x_2\mathbf{i}+x_3\mathbf{j}+x_4\mathbf{k})$$
$$=-2x_1x_2+(x_1^2-x_2^2+x_3^2+x_4^2)\,\mathbf{i}-2x_2x_3\mathbf{j}-2x_2x_4\mathbf{k}$$
$$=(-2x_2x_1,1-2x_2^2,-2x_2x_3,-2x_2x_4),$$
$$=(0,1,0,0)\,\|\delta_{ij}-2x_ix_j\|.$$

Thus the maps t' and h coincide, so that h represents a generator of the kernel of $i_3:\pi_3(\Omega_4)\to\pi_3(\Omega_5)$, whence the kernel is generated by $-\rho+2\sigma$. It is now obvious that $\pi_3(\Omega_5)$ is cyclic infinite, generated by $i_3\sigma$.

THEOREM 3·18. $\pi_3(\Omega_n)=Z_\infty$, $n\geqslant 5$, *generated by the image of* σ.

We will return later to further calculations of homotopy groups of rotation groups.

4. Appendix on pseudo-fibre-spaces.† We have seen that the fundamental theorems on fibre-spaces are theorems 1·1 and 1·2. The point of view adopted by J.-P. Serre and others is that a fibre-space should be defined in terms of these properties. Thus we make the definition that X is a *pseudo-fibre-space* over Y with projection ϕ if there is a map $\phi: X \to Y$, of X on to Y such that theorem 1·2 holds.

THEOREM 4·1. *If X is a pseudo-fibre-space over Y with projection ϕ then the conclusion of theorem* 1·1 *holds.*

We adopt the notation of theorem 1·2 and prove first that the lifting homotopy property of the map $\phi: X \to Y$ implies that if L is a subcomplex of K and if $g_t: K \to Y$ is a homotopy rel L then we may choose the covering homotopy $f_t: K \to Y$ also to be a homotopy rel L. For we proceed step-by-step up the sections of K. Suppose f_t has been defined on K^n so that $f_t \mid L^n = f \mid L^n$ and let $\sigma \in K^{n+1}$. If $\sigma \in L$ we define $f_t \mid \sigma = f \mid \sigma$. Now for any closed simplex σ it is easy to see that the pair $\sigma \times I, \sigma \times 0 \cup \dot{\sigma} \times I$ is homeomorphic to the pair $\sigma \times I, \sigma \times 0$. Thus the lifting homotopy property for ϕ ensures that, if $\sigma \in K - L$, we may extend $f_t \mid \dot{\sigma}$ to σ covering $g_t \mid \sigma$. In this way we extend f_t over the whole of K^{n+1}; and the homotopy $f_t: K \to X$ which finally emerges is plainly stationary on L.

We now adopt the notation of theorem 1·1. We thus have (see p. 48) a homotopy $\rho: K \times I \to K$ with $\rho(p, 1) = p, \rho(q, t) = q$, $\rho(p, 0) \in L, p \in K, q \in L$. We define

$$f'_0: K \to X, \quad g'_t: K \to Y,$$

by $$f'_0(p) = f_0\rho(p, 0), \quad g'_t(p) = g\rho(p, t).$$

Then $f'_0(p) = g_0\rho(p, 0) = g'_0(p)$ and $g'_t(q) = g(q)$, so that g'_t is rel L. It follows therefore that we may choose $f'_t: K \to X$, covering g'_t, to be a homotopy rel L. We define $f: K \to Y$ by $f = f'_1$. Then $f(q) = f'_1(q) = f'_0(q) = f_0(q)$, so that f extends f_0; and $\phi f(p) = \phi f'_1(p) = g'_1(p) = g\rho(p, 1) = g(p)$ so that f covers g. This completes the proof of theorem 4·1.

We remark that we have already demonstrated the trivial

† This section may be omitted on first reading.

fact that the conclusion of theorem 1·1 implies that of theorem 1·2. Thus theorem 4·1 serves to establish that the conclusions of those two theorems are logically equivalent; either may be taken as defining a pseudo-fibre-space.

Thus we may say that the triple (X, Y, ϕ) is a *pseudo-fibration* if ϕ has the lifting homotopy property with respect to maps of finite simplical complexes; and we have proved that an equivalent definition of a pseudo-fibration would be that the conclusion of theorem 1·1 holds for the triple (X, Y, ϕ). However, the proof of theorem 4·1 shows that, in defining pseudo-fibre-spaces, it is sufficient to require that $\phi: X \to Y$ have the lifting homotopy property for maps of (closed) simplexes, or, equivalently, of cubes. For the first half of the proof of theorem 4·1 is effectively a demonstration of the fact that, if ϕ has the lifting homotopy property for maps of simplexes then it has the lifting homotopy property for maps of finite simplical complexes, and moreover that the lifted homotopy may be taken to be stationary on any subcomplex on which the given homotopy was stationary. Indeed, Serre, in his fundamental paper, Homologie Singulière des Espaces Fibrés, *Ann. Math.* 54 (1951), 425–505, defined a *fibre-space* precisely by the lifting homotopy property for maps of cubes. What we have defined as a fibre-space appears in his terminology as a *locally trivial* fibre-space; locally trivial, that is, in the sense that the fibre-space X looks locally like the topological product of the base and the fibre A.

We may further remark that the projection ϕ of a pseudo-fibre-space has the lifting homotopy property even for maps of infinite complexes.†

Let us call the sets $\phi^{-1}(y)$, $y \in Y$, the fibres (or pseudo-fibres). Then if A_0 is the fibre over $y_0 \in Y$, and if ϕ induces

$$\phi_r : \pi_r(X, A_0) \to \pi_r(Y),$$

then theorem 1·3 holds, so that $\phi_r : \pi_r(X, A_0) \approx \pi_r(Y)$. If i_n, j_n, d_n are the homomorphisms of the homotopy sequence of the pair (X, A_0), and if $\bar{d}_{n+1} = d_{n+1}\phi_{n+1}^{-1}$, $\bar{j}_n = \phi_n j_n$, then we get the exact

† For the first half of the proof of theorem 4·1 does not require that K be finite.

sequence

$$\ldots \longrightarrow \pi_{n+1}(Y) \xrightarrow{\bar{d}_{n+1}} \pi_n(A_0) \xrightarrow{i_n} \pi_n(X) \xrightarrow{\bar{j}_n} \pi_n(Y) \rightarrow \ldots.$$

The importance of the generalization to pseudo-fibre-spaces lies in their application to mapping-spaces.

Given two arcwise-connected spaces X and Y, let Y^X stand for the space of maps $f: X \rightarrow Y$, with the compact-open topology.† Let $E_{A,B}$ be the subspace of Y^I consisting of maps $f: I \rightarrow Y$ with $f(0) \subset A$, $f(1) \subset B$ (i.e. the space of paths beginning in A and ending in B). Then $E_{y,y}$ is the space of closed paths on Y with base point y. We write Z_y for $E_{y,y}$. We define the projection $\phi: E_{A,B} \rightarrow A \times B$ by $\phi(f) = (f(0), f(1))$.

THEOREM 4·2. *$E_{A,B}$ is a pseudo-fibre-space over $A \times B$ with projection ϕ.*

We have to show that if f_0 is a map of a (finite) simplicial complex K into $E_{A,B}$ and if $g_0 = \phi f_0: K \rightarrow A \times B$ admits the homotopy $g_t: K \rightarrow A \times B$, then f_0 admits a homotopy $f_t: K \rightarrow E_{A,B}$ such that $g_t = \phi f_t$. In fact, we will show that this statement remains true if K is replaced by an arbitrary space P.

Let P be an arbitrary space, let f_0 be a map $P \rightarrow E_{A,B}$ and let g be a map $P \times I \rightarrow A \times B$, such that $\phi f_0(x) = g(x, 0)$, $x \in P$. Let $g(x, t) = (g_A(x, t), g_B(x, t))$, so that g_A, g_B are maps $P \times I \rightarrow A$, $P \times I \rightarrow B$ respectively. Let $h_0: P \times I \rightarrow Y$ be the map given by $h_0(x, t) = (f_0(x))(t)$, $x \in P$, $t \in I$. Then, by definition,

$$h_0(x, 0) = g_A(x, 0), \quad h_0(x, 1) = g_B(x, 0).$$

We require to find a map $f: P \times I \rightarrow E_{A,B}$ such that $f(x, 0) = f_0(x)$, $\phi f(x, t) = g(x, t)$. This is equivalent to finding a map

$$h: P \times I \times I \rightarrow Y$$

such that $$h(x, t, 0) = h_0(x, t),$$

† Let C be the set of all compact subsets of X, G the set of all open subsets of Y. Then an element of a sub-base of open sets of Y^X is a set, U, of all maps, f, of X into Y such that $f(C) \subset G$ for a particular $C \in$ C and a particular $G \in$ G. The essential property we require on the topology of Y^X is that if $\phi: Z \rightarrow Y^X$, $\psi: Z \times X \rightarrow Y$ are transformations related by $(\phi(z))(x) = \psi(z, x)$, $z \in Z$, $x \in X$, then ϕ is continuous if and only if ψ is continuous. This is true if Y^X has the compact-open topology and X is locally compact.

$$h(x,0,t)=g_A(x,t),$$

$$h(x,1,t)=g_B(x,t).$$

Since $I\times 0\cup 0\times I\cup 1\times I$ is a retract of $I\times I$, it follows that such a map h may be found, and the theorem is proved.

THEOREM 4·3. *If A may be deformed over Y to a point y, then* $E_{A,B}\sim A\times E_{y,B}$.

Let $F:A\times I\to Y$ be a map such that $F(a,0)=a$, $F(a,1)=y$, $a\in A$. Let f_a be the path given by $f_a(t)=F(a,t)$, f'_a the path given by $f'_a(t)=F(a,1-t)$. Let $\phi:A\times E_{y,B}\to E_{A,B}$ be given by† $\phi(a,f)=f_a+f$, and let $\psi:E_{A,B}\to A\times E_{y,B}$ be given by

$$\psi(g)=(a,f'_a+g),$$

where $a=g(0)$. Then

$$\phi\psi(g)=f_a+f'_a+g,\quad a=g(0),\quad \psi\phi(a,f)=(a,f'_a+f_a+f).$$

Since $f_a+f'_a$ and f'_a+f_a are homotopic to constant maps, keeping the image of 0 fixed (in the one case at a, in the other at y), it follows that $\phi\psi\sim 1$, $\psi\phi\sim 1$, and the theorem is proved.

THEOREM 4·4. *The homotopy type of $E_{y,y'}$ is independent of the choice of $y,y'\in Y$.*

For if y_1,y'_1 are any other two points of Y, $E_{y,y'}\sim E_{y_1,y'}\sim E_{y_1,y'_1}$, since y is deformable to y_1 (and y' to y'_1), Y being arcwise-connected.

It follows from this that the homology and homotopy groups of the $E_{y,y'}$ are all isomorphic, and are, in particular, isomorphic to those of Z_y. It should be noted that Z_y is arcwise-connected if and only if Y is simply-connected. We fix a base-point $y_0\in Y$ and write Z for Z_{y_0}. We then get a pseudo-fibre-space $E_{y_0,Y}$ with base-space Y and 'fibre' Z. We note, moreover, that the space $E_{y_0,Y}$ is contractible, and so has trivial homology and homotopy groups. Applying the homotopy sequence simply gives the obvious result $\pi_n(Z)\approx\pi_{n+1}(Y)$. However, it is possible to obtain information about the homology groups of Z, knowing those of Y.

Now let Y be simply-connected and let n be greater than 1. Write $Y=Y_0$, Z_0 the space of closed paths on Y_0, Y_1 the universal‡

† The sum of two maps is defined as in Chapter II.

§ We take universal covers so that all the Z_r may be arcwise-connected.

cover of Z_0, Z_1 the space of closed paths on $Y_1, \ldots, Y_{r+1}$ the universal cover of Z_r, Z_{r+1} the space of closed paths on $Y_{r+1}, \ldots$.

Then

$$\pi_n(Y) \approx \pi_{n-1}(Z_0) \approx \pi_{n-1}(Y_1) \approx \pi_{n-2}(Z_1) \approx \ldots \approx \pi_1(Z_{n-2}).$$

Also, since $\pi_n(Y)$ is Abelian, $\pi_1(Z_{n-2}) \approx H_1(Z_{n-2})$. Thus $\pi_n(Y)$ is represented as the 1-dimensional homology group of a suitable space of closed paths.

The device of taking the universal cover of a space X has the useful effect of 'killing' the fundamental group of X, i.e. we replace X by a space which has the same homotopy groups as X in dimensions $2, 3, \ldots$, but whose fundamental group is zero. We may generalize the operation of going over to the universal covering space as follows.

Let X be an arcwise-connected space† such that $\pi_n(X) = 0$, $n = 1, \ldots, m-1$. By attaching cells of dimension $(m+2)$ corresponding to generators of $\pi_{m+1}(X)$, we may embed X in a space X_1 such that $\pi_n(X_1) = \pi_n(X)$, $n = 1, 2, \ldots, m$, $\pi_{m+1}(X_1) = 0$. We may now attach $(m+3)$-cells to X_1 to 'kill' $\pi_{m+2}(X_1)$ and we may proceed in this way until we reach a space Y, containing X, and such that $\pi_n(Y) = 0$, $n \neq m$, $\pi_m(Y) = \pi_m(X)$. Let E be the space of paths on Y beginning in X and ending in x_0 $(E = E_{X, x_0})$, and let $\phi: E \to X$ be the map which associates with such a path its initial point. Then, just as before, E is a pseudo-fibre-space over X with 'fibres' of the homotopy type of Z, the space of closed paths on Y.

THEOREM 4·5. *The homotopy groups of E are given by*

$$\pi_n(E) \approx \pi_n(X), \quad n \geq m+1, \quad \pi_n(E) = 0, \quad n < m+1.$$

Consider the sequence

$$\ldots \to \pi_{n+1}(X) \xrightarrow{d_{n+1}} \pi_n(Z) \xrightarrow{i} \pi_n(E) \xrightarrow{j} \pi_n(X) \xrightarrow{d_n} \pi_{n-1}(Z) \to \ldots$$

If $n \geq m+1$, $\pi_n(Z) = \pi_{n+1}(Y) = 0$, $\pi_{n-1}(Z) = \pi_n(Y) = 0$, and

$$j: \pi_n(E) \approx \pi_n(X).$$

If $n < m-1$, $\pi_n(Z) = \pi_{n-1}(Z) = 0$ and $\pi_n(E) \approx \pi_n(X) = 0$.

† For a precise definition of 'attaching a cell to a space', see § 2 of Chapter VI.

Now consider the homomorphism $\bar{d}_m : \pi_m(X) \to \pi_{m-1}(Z)$. It is induced as follows. If $f : I^m, \dot{I}^m \to X, x_0$ represents $\alpha \in \pi_m(X)$, then $\bar{d}_m \alpha$ is represented by $g : I^{m-1}, \dot{I}^{m-1} \to Z, z_0$, where

$$(g(t_1, \ldots, t_{m-1}))\,(t_m) = f(t_1, \ldots, t_{m-1}, t_m),$$

and z_0 is the constant path at x_0. It is thus clear that if i'_m is the injection $\pi_m(X) \to \pi_m(Y)$, and $\bar{d}'_m$ the usual isomorphism† $\pi_m(Y) \approx \pi_{m-1}(Z)$, then $\bar{d}_m = \bar{d}'_m i'_m$. Since i'_m is an isomorphism of $\pi_m(X)$ on to $\pi_m(Y)$, it follows that $\bar{d}_m$ is an isomorphism of $\pi_m(X)$ on to $\pi_{m-1}(Z)$. Put $n = m$. Then $\pi_n(Z) = 0$, so that $\bar{\jmath}$ is an isomorphism of $\pi_n(E)$ into $\pi_n(X)$, and $\bar{d}_n$ is an isomorphism, so that $\bar{\jmath}$ maps $\pi_n(E)$ to zero. Thus $\pi_m(E) = 0$. Put $n = m - 1$. Then $\pi_{n-1}(Z) = 0$, so that $\bar{\jmath}$ maps $\pi_n(E)$ on to $\pi_n(X)$, and $\bar{d}_{n+1}$ maps $\pi_{n+1}(X)$ on to $\pi_n(Z)$, so that i maps $\pi_n(Z)$ to 0 and $\bar{\jmath}$ is an isomorphism. Thus $\bar{\jmath} : \pi_{m-1}(E) \approx \pi_{m-1}(X) = 0$, and the theorem is proved.

We have actually shown that the isomorphism between $\pi_n(E)$ and $\pi_n(X)$ is $\bar{\jmath}$, which is induced by $\phi : E \to X$. The 'fibre' Z is a space all of whose homotopy groups vanish except $\pi_{m-1}(Z)$, which is isomorphic to $\pi_m(X)$. Such a space is a geometrical realization of the Eilenberg-Maclane‡ complex $K(\pi_m(X), m-1)$. Its homology groups have been extensively studied.

By the Hurewicz isomorphism theorem, $\pi_{m+1}(E) \approx H_{m+1}(E)$. Thus $\pi_{m+1}(X) \approx H_{m+1}(E)$, so that if we know the homology groups of E we know $\pi_{m+1}(X)$. If we repeat the process starting with E, we may identify $\pi_{m+2}(X)$ as a certain homology group, and so on. Thus we have here an alternative procedure for studying the homotopy groups of spaces. However, they require a deep analysis of the homology relations between pseudo-fibre-spaces, base-spaces, and 'fibres', and this would form the subject of another book.

† The isomorphism $\bar{d}'_n : \pi_n(Y) \approx \pi_{n-1}(Z)$ holds, of course, for all $n \geqslant 2$. It also holds for $n = 1$ if $\pi_0(Z)$ is interpreted as the collection of arcwise-components of Z, and $\bar{d}'_1$ is then (1-1).

‡ See Eilenberg, S., and Maclane, S., *Ann. Math.* 46 (1945), 480–509; ibid. 51 (1950), 514–33.

CHAPTER VI

THE HOPF INVARIANT AND SUSPENSION THEOREMS

1. The Hopf invariant. Let f be a simplicial map of S^{2n-1} on to S^n with respect to given triangulations of S^{2n-1} and S^n, and let p be an inner point of an n-simplex of S^n. Then $f^{-1}(p)$ is an $(n-1)$-chain on some subdivision of the triangulation of S^{2n-1}, and it follows from the fact that S^{2n-1} is a closed manifold that $f^{-1}(p)$ is, in fact, a cycle. This cycle bounds an n-chain C^n on this subdivision, and by subdividing the given triangulation of S^n suitably, it can be arranged that f is simplicial† on C^n. Then $f(C^n)$ is an n-cycle on S^n, so that, if we call the fundamental n-cycle on S^n by the same symbol S^n, $f(C^n) = \gamma S^n$, for some integer γ. This integer γ is the *Hopf invariant* of the map f. It was first defined by Hopf in his paper, 'Über die Abbildungen der dreidimensionalen Sphäre auf die Kügelfläche', *Math. Ann.* 104 (1931), 637–65, in which, as the title indicates, only maps of S^3 into S^2 were considered, and then, more generally, in his later paper, 'Über die Abbildungen von Sphären auf Sphären niedrigerer Dimension', *Fundam. Math.* 25 (1935), 427–40. We refer the reader to these papers for the elegant proofs of many of the results mentioned in this section.‡

Let q be any other point interior to an n-simplex of S^n (which we may assume different from that containing p). We note first that γ may be characterized as the (algebraic) number of times

† As in Chapter III, we allow f to stand both for a simplicial map and the induced chain-mapping. We also allow C^n to stand for a chain and the subcomplex carrying it, if no confusion will arise.

‡ Although many of the arguments may be simplified thereby, we do not give the cohomology interpretation of the Hopf invariant because we wish to encourage the reader to consult Hopf's original papers and because we do not wish to assume familiarity with cohomology theory. For this interpretation, see Steenrod, 'Cohomology invariants of mappings', *Ann. Math.* 50 (1949), 954–88.

$f(C^n)$ covers q, that is to say, it is the intersection number $\mathscr{I}(C^n, f^{-1}(q))$. This is, by definition, the linking number

$$\mathscr{L}(f^{-1}(p), f^{-1}(q)).$$

If we write γ_p for the Hopf invariant of the map f defined with respect to the point p, then

$$\gamma_p = \mathscr{L}(f^{-1}(p), f^{-1}(q)).$$

Now, by a fundamental theorem on linking numbers,†

$$\mathscr{L}(f^{-1}(p), f^{-1}(q)) = (-1)^n \mathscr{L}(f^{-1}(q), f^{-1}(p)).$$

Thus $\gamma_p = (-1)^n \gamma_q$. If r is another point of S^n, then $\gamma_q = (-1)^n \gamma_r$, $\gamma_p = (-1)^n \gamma_r$. Thus, if n is even, γ_p is independent of the choice of p; but, if n is odd, $\gamma_p = 0$. Thus the Hopf invariant is always zero if n is odd, and we restrict ourselves for the rest of this section to the cases in which n is even.

Now let P, Q be given triangulations of S^{2n-1}, S^n and let $f: S^{2n-1} \to S^n$ be simplicial with respect to the triangulations P, Q. We describe $f': S^{2n-1} \to S^n$ as a *simplicial approximation* to f if f' is simplicial with respect to subdivisions P', Q' of P, Q and if, for each vertex σ^0 of S^{2n-1} in the triangulation P',

$$f(\text{star of } \sigma^0) \subset \text{star of } f'(\sigma^0).$$

The stars are, of course, to be taken with respect to the triangulations P', Q'. Hopf showed that if f' is a simplicial approximation to f, then $\gamma(f') = \gamma(f)$, and then extended this result to maps $f_1, f_2: S^{2n-1} \to S^n$ which were homotopic and simplicial with respect to triangulations P_1, P_2 of S^{2n-1} and Q_1, Q_2 of S^n respectively, such that P_1, P_2 have a common subdivision and Q_1, Q_2 have a common subdivision. However, since it is not known whether, in general, two triangulations of a sphere have a common subdivision, this does not yet establish that γ is an invariant of homotopy class. To prove this, he used the following two lemmas, which should be contrasted with each other:

LEMMA 1·1. *Let $g: S_1^{2n-1} \to S^{2n-1}$ be a simplicial map of degree d, $f: S^{2n-1} \to S^n$ a simplicial map. Then $\gamma(fg) = d\gamma(f)$.*

† See Seifert and Threlfall, *Lehrbuch der Topologie*, §77.

For let p be interior to an n-simplex σ^n of S^n. Then $(fg)^{-1}(p)$ consists of a number of $(n-1)$-circuits, each of which is mapped by g on to an $(n-1)$-circuit of $f^{-1}(p)$. If m_1 $(n-1)$-circuits, of $(fg)^{-1}(p)$, are mapped positively and m_2 $(n-1)$-circuits, of $(fg)^{-1}(p)$, are mapped negatively by g on to a particular $(n-1)$-circuit of $f^{-1}(p)$, then $m_1-m_2=d$, so that, regarding f and g as chain-mappings, $g((fg)^{-1}(p))=d(f^{-1}(p))$. Let $(fg)^{-1}(p)$ bound C_1^n in S_1^{2n-1} and $f^{-1}(p)$ bound C^n in S^{2n-1}. Then† $g(C_1^n)-dC^n$ is an n-cycle of S^{2n-1}, and hence a bounding n-cycle. It follows that $fg(C_1^n)-df(C^n)$ is a bounding n-cycle of S^n, and hence zero. Thus $\gamma(fg)=d\gamma(f)$.

LEMMA 1·2. *Let $g: S^n \to S_1^n$ be a simplicial map of degree d, $f: S^{2n-1} \to S^n$ a simplicial map. Then $\gamma(gf)=d^2\gamma(f)$.*

For let p be interior to an n-simplex σ^n of S_1^n. Then m_1 simplexes $\sigma_1^n, \ldots, \sigma_{m_1}^n$ are mapped, by g, positively on σ^n, and m_2 simplexes $\sigma_{m_1+1}^n, \ldots, \sigma_{m_1+m_2}^n$ are mapped, by g, negatively on σ^n, where $m_1-m_2=d$. Let p_i be the point interior to σ_i^n mapped by g on p, $i=1, \ldots, m_1+m_2$. Then, taking account of orientations, $(gf)^{-1}(p)=\sum_{i=1}^{m_1} f^{-1}(p_i)-\sum_{j=1}^{m_2} f^{-1}(p_{m_1+j})$. Let C_i^n be a chain of S^{2n-1} bounded by $f^{-1}(p_i)$, $i=1, \ldots, m_1+m_2$. Then $\sum_{i=1}^{m_1} C_i^n-\sum_{j=1}^{m_2} C_{m_1+j}^n$ is bounded by $(gf)^{-1}(p)$, so that

$$\gamma(gf)\,S_1^n=gf\left(\sum_{i=1}^{m_1} C_i^n-\sum_{j=1}^{m_2} C_{m_1+j}^n\right)=g((m_1-m_2)\,\gamma(f)\,S^n)$$
$$=g(d\gamma(f)\,S^n)=d^2\gamma(f)\,S_1^n,$$

proving the lemma.

THEOREM 1·3. *γ is an invariant of the homotopy class of maps $S^{2n-1} \to S^n$.*

Now let P, Q be triangulations of S^{2n-1}, S^n and let $\{P\}, \{Q\}$ be the systems of triangulations consisting of all subdivisions of P, Q respectively. Given any map $f: S^{2n-1} \to S^n$, we may define $\gamma_{P,Q}(f)$ as the Hopf invariant of any map $f': S^{2n-1} \to S^n$ which is

† We subdivide the given triangulations of S_1^{2n-1}, S^{2n-1}, S^n so that g is simplicial on C_1^n and f is simplicial on C^n. A similar remark will apply in the proof of lemma 1·2.

homotopic to f and simplicial† with respect to some $P' \in \{P\}$, $Q' \in \{Q\}$. The theorem is proved if we can show that

$$\gamma_{P,Q}(f) = \gamma_{P_1,Q_1}(f)$$

for any other triangulation systems $\{P_1\}, \{Q_1\}$. We prove first that $\gamma_{P,Q}(f) = \gamma_{P_1,Q}(f)$. We triangulate S_1^{2n-1} by P_1 and we let $g: S_1^{2n-1} \to S^{2n-1}$ be any map of degree d, $f: S^{2n-1} \to S^n$ any map. Then $f \sim f'$, where f' is simplicial with respect to the triangulations $P' \in \{P\}, Q$ of S^{2n-1}, S^n respectively, and $g \sim g': S_1^{2n-1} \to S^{2n-1}$, where g' is simplicial with respect to the triangulations $P_1' \in \{P_1\}$, P' of S_1^{2n-1}, S^{2n-1} respectively. Then $\gamma_{P_1,Q}(fg) = \gamma(f'g') = d\gamma(f')$, by lemma 1·1, $= d\gamma_{P,Q}(f)$.

Now let S_1^{2n-1}, S^{2n-1} be two copies of the same $(2n-1)$-sphere, S_1^{2n-1} being given the triangulation P_1 and S^{2n-1} the triangulation P. Let $g: S_1^{2n-1} \to S^{2n-1}$ be the map which associates points of S_1^{2n-1} and S^{2n-1} which correspond to the same point of the $(2n-1)$-sphere. Then g is a homeomorphism and will have degree $+1$ if we orient S_1^{2n-1}, S^{2n-1} suitably. Then the map $fg: S_1^{2n-1} \to S^n$ is just the map f regarded as associated with the triangulation P_1 rather than P. Thus

$$\gamma_{P_1,Q}(f) = \gamma_{P_1,Q}(fg) = \gamma_{P,Q}(f),$$

by the first half of the proof. The proof that $\gamma_{P_1,Q}(f) = \gamma_{P_1,Q_1}(f)$ follows similar lines, using lemma 1·2 instead of lemma 1·1. We suppress the details.

Having now defined $\gamma(f)$ for any map $f: S^{2n-1} \to S^n$, we may restate lemmas 1·1 and 1·2 as follows.

LEMMA 1·1*. *Let $g: S_1^{2n-1} \to S^{2n-1}$ be a map of degree d, and $f: S^{2n-1} \to S^n$ any map. Then $\gamma(fg) = d\gamma(f)$.*

LEMMA 1·2*. *Let $g: S^n \to S_1^n$ be a map of degree d, and $f: S^{2n-1} \to S^n$ any map. Then $\gamma(gf) = d^2\gamma(f)$.*

COROLLARY 1·4. *The Hopf invariant of $f: S^{2n-1} \to S^n$ is not affected by changing the orientation of S^n; it is multiplied by -1 if the orientation of S^{2n-1} is reversed.*

† The uniqueness of this definition of $\gamma_{P,Q}(f)$ follows from the paragraph preceding the statement of lemma 1·1.

THEOREM 1·5. *Let* $f: S^{2n-1}, p_0 \to S^n, p_0$ *represent*† $\alpha \in \pi_{2n-1}(S^n)$. *Then* $\alpha \to \gamma(f)$ *defines a homomorphism of* $\pi_{2n-1}(S^n)$ *into the integers.*

Now let $\alpha, \beta \in \pi_{2n-1}(S^n)$ and choose f, g representing α, β so that (i) f, g are simplicial with respect to some triangulation of S^{2n-1}, S^n, (ii) $f(E_-^{2n-1}) = g(E_+^{2n-1}) = p_0$, where E_+^{2n-1}, E_-^{2n-1} are northern and southern hemispheres of S^{2n-1}, and (iii) p_0 is a vertex of $S^n, p_0 = (1, 0, \ldots, 0)$. Then $h: S^{2n-1} \to S^n$, given by $h \mid E_+^{2n-1} = f$, $h \mid E_-^{2n-1} = g$ represents $\alpha + \beta$ and is simplicial.

Let q be interior to an n-simplex of S^n. Then

$$h^{-1}(q) = f^{-1}(q) + g^{-1}(q).$$

Since $f^{-1}(q) \subset E_+^{2n-1}$, it bounds $C_+^n \subset E_+^{2n-1}$. Similarly, $g^{-1}(q)$ bounds $C_-^n \subset E_-^{2n-1}$. Thus

$$\gamma(h)\, S^n = h(C_+^n + C_-^n) = f(C_+^n) + g(C_-^n) = \gamma(f)\, S^n + \gamma(g)\, S^n,$$

or

$$\gamma(h) = \gamma(f) + \gamma(g).$$

It is convenient to write $\gamma(\alpha)$, $\alpha \in \pi_{2n-1}(S^n)$, for the common value of $\gamma(f)$ for all $f: S^{2n-1} \to S^n$ in the class α. Thus $\gamma: \alpha \to \gamma(\alpha)$ is a homomorphism of $\pi_{2n-1}(S^n)$ into Z, the additive group of integers.

We have seen that $\gamma(\alpha) = 0$ if n is odd. The following theorem establishes the non-triviality of γ if n is even:

THEOREM 1·6. *If* n *is even,* $\pi_{2n-1}(S^n)$ *has an element with Hopf invariant* 2.

It follows from corollary 4 that it is sufficient to establish the existence of an element with Hopf invariant ± 2.

Now let E^n be chosen with a definite orientation and let $\phi_n: E^n, S^{n-1} \to S^n, p_0$ be the map defined in §2 of Chapter II, representing the positive generator of $\pi_n(S^n)$. Now S^{2n-1} may be represented as $(E^n \times E^n)^{\cdot}$, and the fundamental $(2n-1)$-cycle is then $\dot{E}^n \times E^n + E^n \times \dot{E}^n$, since n is even. Consider the map $f: S^{2n-1} \to S^n$ given by $f(x, y) = \phi_n(x)$, $y \in \dot{E}^n$, $= \phi_n(y)$, $x \in \dot{E}^n$. We may assume f and ϕ_n simplicial, and recall that ϕ_n is homeomor-

† We may always suppose $S^n \subset S^r$ if $n < r$ by regarding Euclidean $(n+1)$-space as the space spanned by the first $(n+1)$ coordinates of Euclidean $(r+1)$-space.

phic in the interior of E^n. Then, if q is interior to an n-simplex of S^n, the cycle $f^{-1}(q)$ is, except perhaps for a change of orientation, $\dot{E}^n \times p + p \times \dot{E}^n$, where $\phi_n(p) = q$. Let p_1 be any vertex of $\dot{E}^n$ and let C^1 be a chain on a subdivision of E^n bounded by $p - p_1$. Then $\dot{E}^n \times p + p \times \dot{E}^n$ bounds $E^n \times p_1 - \dot{E}^n \times C^1 + p_1 \times E^n + C^1 \times \dot{E}^n$. Since $n > 1$,

$$f(\dot{E}^n \times C^1) \quad \text{and} \quad f(C^1 \times \dot{E}^n)$$

are zero as n-chains of S^n, so that the degree of

$$f \mid (E^n \times p_1 + \dot{E}^n \times C^1 + p_1 \times E^n + C^1 \times \dot{E}^n)$$

is 2 (since $f(E^n \times p_1) = f(p_1 \times E^n) = \phi_n(E^n) = S^n$).

The element described in the proof of this theorem is a special example of a 'Whitehead product', which will be defined in greater generality later in the chapter.

Corollary 1·7. *$\pi_{2n-1}(S^n)$ contains a cyclic infinite subgroup if n is even.*

The subgroup in question is that generated by the element described in theorem 1·6. Corollary 1·7 was proved in Chapter V when $n=2$, 4, or 8. However, the maps described there ($S^3 \to S^2, S^7 \to S^4, S^{15} \to S^8$) all have Hopf invariant ± 1 (according to orientation). This follows from the fact that any two great circles on S^3 have linking number ± 1, and similarly for great 3-spheres on S^7 and great 7-spheres on S^{15}. Thus the projections of $\pi_3(S^2)$ on $\pi_3(S^3)$, of $\pi_7(S^4)$ on $\pi_7(S^7)$, and of $\pi_{15}(S^8)$ on $\pi_{15}(S^{15})$ may be identified with the homomorphism γ.

It is to be noted that, in contrast to the Brouwer degree, we do not assert that two maps $S^{2n-1} \to S^n$ are homotopic if they have the same Hopf invariant. This does happen to be the case if $n=2$, since $\pi_3(S^2) = Z_\infty$, but it is certainly not the case if $n=4$ or 8. It is a curious consequence that, for example, if we follow any map $S^3 \to S^2$ by the antipodal map $S^2 \to S^2$, the composite map is homotopic to the original.

The question arises of whether it is possible to generalize the Hopf invariant. This was first done by G. W. Whitehead, who defined a homomorphism $H: \pi_r(S^n) \to \pi_r(S^{2n-1})$, $r < 3n-3$ (later extended by Blakers and Massey to $r \leqslant 3n-3$). If $r = 2n-1$,

H is a homomorphism of $\pi_{2n-1}(S^n)$ into $\pi_{2n-1}(S^{2n-1})$, and becomes a homomorphism into the integers if each class in $\pi_{2n-1}(S^{2n-1})$ is identified with its characteristic degree. Later, the author defined a homomorphism $H^*: \pi_r(S^n) \to \pi_{r+1}(S^{2n})$, valid for all r, n, which is essentially equivalent (in a sense to be made precise in the next section) to G. W. Whitehead's H if $r \leqslant 3n-3$.

Let $\mu: \pi_r(S^n) \to \pi_r(S^n \cup S^n)$ be the homomorphism induced by that map $S^n \to S^n \cup S^n$ ($S^n \cup S^n$ being the union of two spheres with a single common point), which pinches an equatorial $S^{n-1} \subset S^n$ to a point. Now, by theorem 6·2 of Chapter IV,

$$\pi_r(S^n \cup S^n) \approx \pi_r(S^n) + \pi_r(S^n) + \pi_{r+1}(S^n \times S^n, S^n \cup S^n),$$

where the projections† of $\pi_r(S^n \cup S^n)$ on to its direct summands are explicitly defined. Let Q be the projection

$$\pi_r(S^n \cup S^n) \to \pi_{r+1}(S^n \times S^n, S^n \cup S^n).$$

Finally, let $\chi: \pi_{r+1}(S^n \times S^n, S^n \cup S^n) \to \pi_{r+1}(S^{2n})$ be induced by shrinking $S^n \cup S^n$ to a point. Then we define

$$H^*: \pi_r(S^n) \to \pi_{r+1}(S^{2n}) \quad \text{by} \quad H^* = \chi Q \mu.$$

It will be one of the objectives of this chapter to prove that H^* does generalize the Hopf invariant.

2. The Freudenthal suspension and its generalization. Let f be a map $S^{r-1} \to S^{n-1}$; we may regard S^{r-1}, S^{n-1} as equators of S^r, S^n respectively, and extend f to a map $f': S^r \to S^n$ mapping the northern hemisphere of S^r into the northern hemisphere of S^n, and the southern hemisphere of S^r into the southern hemisphere of S^n. The homotopy class of f' does not depend on what choice of extension of f is made, and, in fact, clearly depends only on the homotopy class of f. Thus we get a mapping

$$F: \pi_{r-1}(S^{n-1}) \to \pi_r(S^n).$$

The map f' is called the *suspension* of f, and, if $\alpha \in \pi_{r-1}(S^{n-1})$, $F(\alpha)$ is called the *suspension* of α.

† See also §4 of this chapter. Note that the notations used here are not the same as those of §6 of Chapter IV.

THEOREM 2·1. *F is a homomorphism.*

We prove this indirectly. The homomorphism

$$d:\pi_r(E^n_+, S^{n-1})\to\pi_{r-1}(S^{n-1})$$

is an isomorphism on to $\pi_{r-1}(S^{n-1})$. In fact, if $f:S^{r-1}\to S^{n-1}$ represents $\alpha\in\pi_{r-1}(S^{n-1})$, then $d^{-1}(\alpha)$ is represented by $f'\,|\,E^r_+$, where E^r_+ is the northern hemisphere of S^r. Let

$$\phi_n: E^n_+, S^{n-1}\to S^n, p_0$$

induce $\phi:\pi_r(E^n_+, S^{n-1})\to\pi_r(S^n)$. Then $\phi_n f'\,|\,E^r_+$ represents $\phi d^{-1}(\alpha)$. We may extend $\phi_n f'\,|\,E^r_+$ to S^r by defining $\phi_n f'(E^r_-)=p_0$. The resulting map still represents $\phi d^{-1}(\alpha)$, by theorem 2·5 of Chapter III, and coincides with $\overline{\phi}_n f'$. Since $\overline{\phi}_n\sim 1: S^n, p_0\to S^n, p_0$, it follows that $\overline{\phi}_n f'$ represents the same element as f', i.e. $F(\alpha)$. Thus $F=\phi d^{-1}$ and F is a homomorphism.

It should be noted that the homomorphism ϕ is equal to $j^{-1}i$, where i is the injection $\pi_r(E^n_+, S^{n-1})\to\pi_r(S^n, E^n_-)$, and j is the injection $\pi_r(S^n)\to\pi_r(S^n, E^n_-)$, which is an isomorphism on to $\pi_r(S^n, E^n_-)$ since E^n_- is contractible over itself to p_0. It is only necessary to observe that $\overline{\phi}_n$ induces $j^{-1}:\pi_r(S^n, E^n_-)\to\pi_r(S^n, p_0)$. Thus $F=\phi d^{-1}=j^{-1}id^{-1}$, and the behaviour of F is referred back to that of i.

Let X be an arcwise-connected Hausdorff space and let e^n be an (open) n-cell, i.e. the homeomorph of the interior of an n-element, disjoint from X. We turn $X\cup e^n$ into a topological space by retaining X as a closed subset of $X\cup e^n$ and defining the closure of e^n as the image of the n-element E^n under a transformation $f: E^n, S^{n-1}\to X\cup e^n, X$, such that $f\,|\,E^n-S^{n-1}$ is a homeomorphism on to e^n. The map f is called a *characteristic map* for e^n, and $f\,|\,S^{n-1}$ an *attaching map* for e^n. We will give the following examples of this fundamental idea:

(A) E^n itself is obtained by attaching the open n-cell (E^n-S^{n-1}) to S^{n-1}.

(B) Let e^2 be attached to S^1 by a map $f:S^1\to S^1$ of degree 2. Then $S^1\cup e^2$ has the homotopy type of the projective plane.

(C) Let $f:E^n\times E^n\to S^n\times S^n$ be given by

$$f(x,y)=(\phi_n(x),\phi_n(y)).$$

Then $f((E^n \times E^n)^{\cdot}) \subset S^n \cup S^n$. Thus f is a map

$$E^n \times E^n, (E^n \times E^n)^{\cdot} \to S^n \times S^n, S^n \cup S^n$$

and it follows from the definition of ϕ_n that

$$f \mid (E^n \times E^n - (E^n \times E^n)^{\cdot})$$

is a homeomorphism on to $S^n \times S^n - (S^n \cup S^n)$. Thus, identifying $E^n \times E^n$ with a $2n$-element, we may say that $S^n \times S^n$ is formed from $S^n \cup S^n$ by attaching a $2n$-cell. The map f is a characteristic map, and $f \mid (E^n \times E^n)^{\cdot}$ is an attaching map.

(D) S^n is formed from E^n_- by attaching the open n-cell $(E^n_+ - S^{n-1})$.

(E) S^n is formed from p_0 by attaching the open n-cell $S^n - p_0$. Then ϕ_n is a characteristic map.

Example (D) shows the relation of the idea of attaching cells to spaces to the idea of suspension. For the homomorphism $i: \pi_r(E^n_+, S^{n-1}) \to \pi_r(S^n, p_0)$ may be regarded† as being induced by the characteristic map for the cell $E^n_+ - S^{n-1}$, in this case the identity map. In general, let $X^* = X \cup e^n$ and let the characteristic map $f: E^n, S^{n-1} \to X^*, X$ induce the homomorphism $g_r: \pi_r(E^n, S^{n-1}) \to \pi_r(X^*, X)$. The homomorphism g_r is called (by J. H. C. Whitehead) the *(generalized) suspension homomorphism*. Recall that $\pi_r(E^n, S^{n-1}) \approx \pi_{r-1}(S^{n-1})$ under the homotopy boundary homomorphism.

Let $\chi_r: \pi_r(X^*, X) \to \pi_r(S^n)$ be induced by shrinking‡ X to a point. Then $\chi_r g_r: \pi_r(E^n, S^{n-1}) \to \pi_r(S^n)$ is induced by a map which is homeomorphic (and orientation-preserving) in $E^n - S^{n-1}$ and maps S^{n-1} to a point. Such a map is homotopic to ϕ_n, so

† Eilenberg has drawn attention to the relation of suspension to 'excision', since (E^n_-, S^{n-1}) is obtained from (S^n, E^n_+) by excising $E^n_+ - S^{n-1}$. The failure of the excision axiom for relative homotopy groups thus distinguishes sharply between homology and homotopy theory (see Eilenberg and Steenrod, 'Axiomatic approach to homology theory', *Proc. Nat. Acad. Sci., Wash.*, 31 (1945), 117–20).

‡ Let $g: E^n, S^{n-1} \to S^n, p_0$ be any map which is an orientation-preserving homeomorphism of $E^n - S^{n-1}$ on to $S^n - p_0$. Then a shrinking map

$$h: X^*, X \to S^n, p_0$$

is given by $h(x) = p_0$, $x \in X$, $h \mid \bar{e}^n = gf^{-1}$, f being the characteristic map for e^n.

that $\chi_r g_r d^{-1}$ is the suspension $F:\pi_{r-1}(S^{n-1})\to\pi_r(S^n)$. We exhibit the relation

$$F_{r-1}=\chi_r g_r d^{-1} \tag{2·2}$$

for future reference, writing F_{r-1} for $F:\pi_{r-1}(S^{n-1})\to\pi_r(S^n)$.

We now prove some elementary theorems on suspension.

THEOREM 2·3. *If* $\alpha\in F\pi_{r-1}(S^{n-1})$, $\beta_1,\beta_2\in\pi_n(X)$, *then*

$$(\beta_1+\beta_2)\circ\alpha=\beta_1\circ\alpha+\beta_2\circ\alpha.$$

Here $\beta\circ\alpha$, $\alpha\in\pi_r(S^n)$, $\beta\in\pi_n(X)$, is that element of $\pi_r(X)$ represented by $gf:S^r\to X$, where $f:S^r\to S^n$ represents α, $g:S^n\to X$ represents β. Represent $\alpha\in F\pi_{r-1}(S^{n-1})$ by $f:S^r\to S^n$, mapping E^r_+ to E^n_+ and E^r_- to E^n_-, and represent $\beta_1,\beta_2\in\pi_n(X)$ by $g_1,g_2:S^n\to X$, such that $g_1(E^n_-)=g_2(E^n_+)=x_0$, the base-point in X. Then let $g:S^n\to X$ be defined by $g\mid E^n_+=g_1$, $g\mid E^n_-=g_2$. The map g represents $\beta_1+\beta_2$, so that gf represents $(\beta_1+\beta_2)\circ\alpha$. Also, g_1f,g_2f represent $\beta_1\circ\alpha,\beta_2\circ\alpha$ respectively. Now $gf\mid E^r_+=g_1f\mid E^r_+$, $gf\mid E^r_-=g_2f\mid E^r_-$ and $g_1f(E^r_-)=g_2f(E^r_+)=x_0$. Thus gf represents $\beta_1\circ\alpha+\beta_2\circ\alpha$ and the theorem is proved.

We might express the conclusion of this theorem by saying that the 'left-distributive law' holds if α is a suspension element (i.e. an element in $F\pi_{r-1}(S^{n-1})$). That the 'right-distributive law' $\beta\circ(\alpha_1+\alpha_2)=\beta\circ\alpha_1+\beta\circ\alpha_2$, $\beta\in\pi_n(X)$, $\alpha_1,\alpha_2\in\pi_r(S^n)$, holds is an immediate consequence of definition, and is just a special case of the fact that a map of X into Y induces a homomorphism of $\pi_r(X)$ into $\pi_r(Y)$. We may show that the 'left-distributive law' does not necessarily hold if $\alpha\in\pi_r(S^n)$ is not a suspension element. For let $\alpha\in\pi_3(S^2)$, $\alpha\neq 0$, and let $1,-1\in\pi_2(S^2)$. Then $(1-1)\circ\alpha=0$, but $(-1)\circ\alpha=\alpha$, so that

$$(-1)\circ\alpha+1\circ\alpha=2\alpha\neq 0, \quad\text{and}\quad (1-1)\circ\alpha\neq 1\circ\alpha+(-1)\circ\alpha.$$

On the other hand, we may have $(\beta_1+\beta_2)\circ\alpha=\beta_1\circ\alpha+\beta_2\circ\alpha$, where $\beta_1\neq 0,\beta_2\neq 0\in\pi_n(X)$ and $\alpha,\in\pi_r(S^n)$, is not a suspension element.†

THEOREM 2·4. *If* $f:E^{r+1},S^r\to E^{n+1},S^n$ *represents* $d^{-1}(\alpha)$, $\alpha\in\pi_r(S^n)$, *where d is the boundary isomorphism*

$$\pi_{r+1}(E^{n+1},S^n)\approx\pi_r(S^n),$$

† This happens if $r=6$, $n=3$ and α is an element of order 3 in $\pi_6(S^3)$. We know that $\pi_6(S^3)$ is cyclic of order 12.

and if $g: E^{n+1}, S^n \to X, x_0$ *represents* $\beta \in \pi_{n+1}(X)$, *then*

$$gf: E^{r+1}, S^r \to X, x_0$$

represents $\beta \circ F\alpha \in \pi_{r+1}(X)$.

For we may replace E^{r+1}, E^{n+1} by E_+^{r+1}, E_+^{n+1} and extend f to $f': S^{r+1} \to S^{n+1}$ by mapping E_-^{r+1} into E_-^{n+1}, and extend g to $g': S^{n+1} \to X$ by mapping E_-^{n+1} to x_0. Then $g'f'$ represents the same element as gf, g' represents the same element as g, namely, β, and f' represents $F\alpha$. Thus $g'f'$, and hence gf, represents $\beta \circ F\alpha$.

THEOREM 2·5. $F(\alpha \circ \beta) = F\alpha \circ F\beta$, $\alpha \in \pi_p(S^n)$, $\beta \in \pi_r(S^p)$.

From this obvious proposition follows the corollary

COROLLARY 2·6. *If* $\alpha \in \pi_r(S^n)$, $-1 \in \pi_n(S^n)$, *then*

$$F((-1) \circ \alpha) = -F\alpha.$$

Since F is a homomorphism $F(-1) = -1 \in \pi_{n+1}(S^{n+1})$. Thus $F((-1) \circ \alpha) = F(-1) \circ F\alpha = -1 \circ F\alpha = -F\alpha$, by theorem 2·3.

COROLLARY 2·7. $F\pi_3(S^2)$ *has at most two elements.*

Let α generate $\pi_3(S^2)$. Then $(-1) \circ \alpha = \alpha$; but

$$F(\alpha + (-1) \circ \alpha) = F\alpha - F\alpha = 0.$$

Thus $F(2\alpha) = 2F(\alpha) = 0$, and the corollary follows.

THEOREM 2·8. *If* $\alpha \in F\pi_{2n-2}(S^{n-1})$, *then the Hopf invariant of* α *is zero.*

Let $f: S^{2n-1}, E_+^{2n-1}, E_-^{2n-1} \to S^n, E_+^n, E_-^n$ be a simplicial map representing α. For any p interior to an n-simplex of E_+^n, $f^{-1}(p)$ is an $(n-1)$-cycle of E_+^{2n-1} which bounds an n-chain C^n in E_+^{2n-1}, since $n \geq 2$. Thus $f(C^n)$ certainly does not cover S^n, so that the Hopf invariant of α is zero.

We may also prove

THEOREM 2·9. *If* $\alpha \in F\pi_{r-1}(S^{n-1})$, *then* $H^*(\alpha) = 0 \in \pi_{r+1}(S^{2n})$.

We recall that $H^* = \chi Q \mu$,

$$\mu: \pi_r(S^n) \to \pi_r(S^n \cup S^n), \quad Q: \pi_r(S^n \cup S^n) \to \pi_{r+1}(S^n \times S^n, S^n \cup S^n),$$

$$\chi: \pi_{r+1}(S^n \times S^n, S^n \cup S^n) \to \pi_{r+1}(S^{2n}).$$

In fact, we prove that $Q\mu(\alpha) = 0$ if α is a suspension element.

Consider the map $f: S^n \to S_1^n \cup S_2^n$, given by $f \mid E_+^n = i_1\phi$, where ϕ is the map ϕ_n of Chapter II, § 2, regarded now as a map $E_+^n \to S_1^n$, and i_1 is the identity map $S_1^n \to S_1^n \cup S_2^n$; and $f \mid E_-^n = i_2\phi'$, similarly defined.† Then f is a map inducing

$$\mu: \pi_r(S^n) \to \pi_r(S_1^n \cup S_2^n).$$

On the other hand, by its definition, f represents

$$(i_1 + i_2) \in \pi_n(S_1^n \cup S_2^n),$$

where i_λ generates $\pi_n(S_\lambda^n)$, regarded as a subgroup of $\pi_n(S_1^n \cup S_2^n)$, $\lambda = 1, 2$. Thus $\mu(\alpha) = (i_1 + i_2) \circ \alpha$. Now if α is a suspension element, it follows from theorem 2·3 that $(i_1 + i_2) \circ \alpha = i_1 \circ \alpha + i_2 \circ \alpha$. Moreover, Q is defined by taking the direct sum decomposition $\pi_r(S_1^n \cup S_2^n) \approx \pi_r(S_1^n) + \pi_r(S_2^n) + \pi_{r+1}(S_1^n \times S_2^n, S_1^n \cup S_2^n)$ and projecting an element of $\pi_r(S_1^n \cup S_2^n)$ on to its component in

$$\pi_{r+1}(S_1^n \times S_2^n, S_1^n \cup S_2^n).$$

Since $i_1 \circ \alpha + i_2 \circ \alpha \in \pi_r(S_1^n) + \pi_r(S_2^n)$, it follows that the projection of $i_1 \circ \alpha + i_2 \circ \alpha$ is zero, or that $Q\mu(\alpha) = 0$.

We now state the fundamental suspension theorems due to Freudenthal.‡ They have been divided by J. H. C. Whitehead into 'crude' and 'delicate' theorems.

THEOREM 2·10. (Crude suspension theorem.)

$$F: \pi_{r-1}(S^{n-1}) \to \pi_r(S^n)$$

is on to $\pi_r(S^n)$ *if* $r \leqslant 2n-2$ *and isomorphic if* $r < 2n-2$.

The first part of this theorem has been generalized by J. H. C. Whitehead as follows (actually, Whitehead's theorem is even more general than this):

THEOREM 2·11. *Let* $X^* = X \cup e^n$ *and let* $f: E^n, S^{n-1} \to X^*, X$ *be a characteristic map for* e^n. *Then if* f *induces*

$$g_r: \pi_r(E^n, S^{n-1}) \to \pi_r(X^*, X),$$

and if $\pi_s(X) = 0$, $s = 1, \ldots, k < n$, *then* g_r *is on to* $\pi_r(X^*, X)$ *if* $2 \leqslant r \leqslant n + k - 1$.

† That is to say, ϕ_n' is a map $E_-^n, S^{n-1} \to S^n, p_0$, which may be taken as $\phi_n\eta$, where $\eta: E_-^n \to E_+^n$ is the reflexion in S^{n-1}. Then, in this context, ϕ' is ϕ_n' regarded as a map $E_-^n \to S_2^n$ and i_2 is the identity map $S_2^n \to S_1^n \cup S_2^n$.

‡ 'Suspension' is the English translation of Freudenthal's *Einhängung*, introduced in his paper 'Über die Klassen von Sphärenabbildungen I', *Comp. Math.* 5 (1937), 299–314, to which reference should be made.

Since theorem 2·10 asserts that $i:\pi_r(E^n_+, S^{n-1}) \to \pi_r(S^n, E^n_-)$ is on to $\pi_r(S^n, E^n_-)$ if $r \leqslant 2n-2$, and since $\pi_s(E^n_-) = 0$, $s = 1, \ldots, n-1$ (trivially), it is clear that theorem 2·11 does generalize theorem 2·10.

We will show that it is sufficient to prove that, in every class $\alpha \in \pi_r(X^*, X)$, we may find a map $q: E^r, S^{r-1} \to X^*, X$ such that, for some $x \in e^n$, $q^{-1}(x)$ is empty or a single point. Let $f^{-1}(x) = y \in E^n$, then a deformation of $E^n - y$ on to S^{n-1} induces a deformation of $X^* - x$ on to X. This means that the injection

$$i: \pi_r(X^*, X) \to \pi_r(X^*, X^* - x)$$

is an isomorphism (on to). If $q^{-1}(x)$ is empty, $i\alpha = 0$, so that $\alpha = 0$. Assume then that $q^{-1}(x) = z$, a point in the interior of E^r. Let U be a neighbourhood of x in e^n. Then there exists a neighbourhood, V, of z, lying in the interior of E^r and such that $q(\bar{V}) \subset U$. A deformation-retraction of E^r on to $\bar{V}$ gives rise to a homotopy of q under which S^{r-1} does not cross x. This means that $i\alpha, \in \pi_r(X^*, X^* - x)$, may be represented by a map q' having the property that $q'E^r \subset U$. Then $f^{-1}q'$ is a map

$$E^r, S^{r-1} \to E^n, E^n - y,$$

and $f(f^{-1}q') = q'$. Thus $i\alpha$ lies in $\dot{h}_r \pi_r(E^n, E^n - y)$, where

$$h_r: \pi_r(E^n, E^n - y) \to \pi_r(X^*, X^* - x)$$

is induced by f. If j is the injection isomorphism

$$\pi_r(E^n, S^{n-1}) \approx \pi_r(E^n, E^n - y),$$

then clearly $h_r j = ig_r$, so that

$$i\alpha = h_r\beta, \quad \beta \in \pi_r(E^n, E^n - y), = h_r j\gamma, \quad \gamma \in \pi_r(E^n, S^{n-1}), = ig_r\gamma,$$

and, since i is isomorphic, $\alpha = g_r\gamma$, so that g_r is on to

The proof that, under the conditions of the theorem, every class $\in \pi_r(X^*, X)$ may be represented by a map of the desired kind, may be found in J. H. C. Whitehead,† 'A note on suspension', *Quart. J. Math.* (Oxford, 1950), pp. 9–22.

For the proof of the second part of theorem 2·10, the reader may refer to Whitehead's paper or to Freudenthal's original paper.

† In this paper Whitehead also proves a theorem which is, in an important respect, weaker than our theorem 2·12.

THEOREM 2·12. *If g_r is defined as in theorem* 2·11, *then*

$$g_r : \pi_r(E^n, S^{n-1}) \to \pi_r(X^*, X)$$

is isomorphic (into) whenever $F : \pi_{r-1}(S^{n-1}) \to \pi_r(S^n)$ is an isomorphism (into). If F is also on to $\pi_r(S^n)$ (e.g. if $r < 2n-2$), g_r is an isomorphism on to a direct summand in $\pi_r(X^*, X)$.

For, by (2·2), $\chi_r g_r(\alpha) = Fd(\alpha)$, for all $\alpha \in \pi_r(E^n, S^{n-1})$. Thus $\chi_r g_r$ is isomorphic if F is isomorphic, since d is an isomorphism (on to). Clearly g_r is isomorphic if $\chi_r g_r$ is isomorphic, and the first part of the theorem follows. The second part is seen to be an immediate consequence of the following algebraical lemma:

LEMMA 2·13. *Let G, H, K be three Abelian groups and let $\lambda : G \to H$, $\mu : H \to K$ be homomorphisms such that $\mu\lambda : G \approx K$, then λ is an isomorphism (into), and $H = \lambda G + \mu^{-1}(0)$.*

For let $h \in H$. Then $\mu h \in K$, so that $\mu h = \mu\lambda g$, for some $g \in G$. Thus $h = \lambda g + h_0$, $h_0 \in \mu^{-1}(0)$. Now let $\lambda g + h_0 = 0$, $g \in G$, $h_0 \in \mu^{-1}(0)$. Then $\mu\lambda g = \mu(\lambda g + h_0) = 0$, so that $g = 0$, whence λ is isomorphic, and also $h_0 = 0$.

In the terminology of theorem 2·12 we would have

$$\pi_r(X^*, X) = g_r \pi_r(E^n, S^{n-1}) + \chi_r^{-1}(0).$$

An important consequence of theorems 2·11 and 2·12 is†

THEOREM 2·14. *If g_r is defined as in theorem* 2·11, *and if $\pi_s(X) = 0$, $s = 1, 2, \ldots, k < n-1$, then $g_r : \pi_r(E^n, S^{n-1}) \approx \pi_r(X^*, X)$, $r = 2, \ldots, n+k-1$.*

Let $[i_n, i_n] \in \pi_{2n-1}(S^n)$ be the 'Whitehead product' element described in the proof of theorem 1·6. Then the 'delicate' suspension theorems are as follows:

THEOREM 2·15. (Delicate suspension theorems.) *Let $\pi_{2n-1}(S^n)_0$ represent the subgroup of $\pi_{2n-1}(S^n)$ consisting of elements with Hopf invariant* 0. *Then* (i) $F\pi_{2n-2}(S^{n-1}) = \pi_{2n-1}(S^n)_0$, (ii) *the kernel of $F : \pi_{2n-1}(S^n) \to \pi_{2n}(S^{n+1})$ is the cyclic group generated by $[i_n, i_n]$; this element has Hopf invariant ± 2 if n is even, so that the kernel of F is cyclic infinite; if n is odd, then $[i_n, i_n]$ is* 0 *or of order* 2 *according as $\pi_{2n+1}(S^{n+1})$ has or has not an element with Hopf invariant* 1.

† This theorem should be compared with theorem 1 of Whitehead's 'A note on suspension' (see previous footnote and the text to which it refers).

Part (ii) of this theorem is a strengthened form (due to G. W. Whitehead) of Freudenthal's original results. Freudenthal showed that $F:\pi_{2n-1}(S^n)\rightarrow\pi_{2n}(S^{n+1})$, n even, is isomorphic on the subgroup $\pi_{2n-1}(S^n)_0$ of $\pi_{2n-1}(S^n)$ and has a kernel all of whose elements have even Hopf invariant; and, if n is odd,

$$F:\pi_{2n-1}(S^n)\rightarrow\pi_{2n}(S^{n+1})$$

is isomorphic if $\pi_{2n+1}(S^{n+1})$ has an element with Hopf invariant 1. This is sufficient to conclude

THEOREM 2·16. *$\pi_{n+1}(S^n)$, $n\geqslant 3$, is cyclic of order 2, generated by $F^{n-2}(\alpha)$, where F^{n-2} is the $(n-2)$-fold suspension and $\alpha\in\pi_3(S^2)$ is the class of the Hopf map.*

For by theorem 2·10,

$$\pi_4(S^3)=F\pi_3(S^2) \quad \text{and} \quad \pi_4(S^3)\approx\pi_5(S^4)\approx\ldots\approx\pi_{n+1}(S^n)\approx\ldots.$$

By corollary 2·7, $\pi_4(S^3)$ has at most two elements, and, by theorem 2·15, $F(\alpha)\neq 0$, since α has Hopf invariant 1.

COROLLARY 2·17. *$\pi_4(S^2)$ is cyclic of order 2, generated by $\alpha\circ F(\alpha)$.*

This follows immediately from theorem 2·1 of Chapter V.

It now follows from theorem 2·15 that $\pi_5(S^3)$ is the homomorphic image of $\pi_4(S^2)$, under F, and that repeated suspension gives isomorphisms $\pi_5(S^3)\approx\pi_6(S^4)\approx\ldots\approx\pi_{n+2}(S^n)\approx\ldots$, since $\pi_7(S^4)$ has an element with Hopf invariant 1. In fact, it has been proved by G. W. Whitehead† that $\pi_{n+2}(S^n)$ is cyclic of order 2, $n\geqslant 2$, generated by $F^{n-2}(\alpha)\circ F^{n-1}(\alpha)$. However, we cannot enter further at this stage on a discussion of the methods of G. W. Whitehead and, more recently, of H. Cartan and J.-P. Serre for calculating higher homotopy groups of spheres. We gather together into one theorem, however, some conclusions which we may draw from the suspension theorems and previous results.

THEOREM 2·18. (i) *$\pi_{n+2}(S^n)=0$ or Z_2, $n\geqslant 3$,* (ii) *$\pi_{n+3}(S^n)$ is constant‡ and non-zero, $n\geqslant 5$,* (iii) *$\pi_{n+7}(S^n)$ is constant and non-*

† See 'On the $(n+2)$nd homotopy group of the n-sphere', *Ann. Math.* 52 (1950), 245–7. This result has also been proved by Pontrjagin, *Doklady Akad. Nauk SSSR*, 70 (1950), 957–9, thus correcting an earlier, erroneous announcement by the same author.

‡ It is now known that $\pi_{n+3}(S^n)$ is cyclic of order 24, $n\geqslant 5$, generated by the $(n-4)$-fold suspension of the Hopf class in $\pi_7(S^4)$.

zero, $n \geqslant 9$, (iv) $\pi_8(S^4)$, $\pi_{10}(S^4)$, $\pi_{16}(S^8)$, $\pi_{18}(S^8)$ *and* $\pi_{22}(S^8)$ *are non-zero.*

Parts (ii) and (iii) follow from theorem 2·15 and the fact that $\pi_7(S^4)$ and $\pi_{15}(S^8)$ have elements with Hopf invariant 1. Part (iv) follows from theorems 2·3 and 2·10 of Chapter V, together with parts (ii) and (iii).

We may display a map $S^n \to S^{n-1}$ representing a generator of $\pi_n(S^{n-1})$, $n \geqslant 3$. Let S^{n-1} be the space of points $(y_1, \ldots, y_n)$ with $y_1^2 + \ldots + y_n^2 = 1$, and S^n be the space of points $(x_1, \ldots, x_{n+1})$ with $x_1^2 + \ldots + x_{n+1}^2 = 1$. Then such a map is given by

$$y_1 = 2\,|\,\xi\,|\,x_1, \ldots, \quad y_{n-3} = 2\,|\,\xi\,|\,x_{n-3}, \quad \eta = 2\xi\zeta, \quad y_n = 1 - 2\,|\,\xi\,|^2,$$

where

$$\xi = x_n + ix_{n+1},\ \eta = y_{n-2} + iy_{n-1},\ \zeta = x_{n-2} + ix_{n-1},\ i = \sqrt{-1}.$$

We have seen that considerable importance attaches to the question of whether there exist elements in $\pi_{2n-1}(S^n)$, n even, with Hopf invariant 1. G. W. Whitehead has shown that this is not the case if $n = 4m + 2$, $m \geqslant 1$, and, more recently,† Adem has proved that, if such elements exist, then $n = 2^k$.

3. Application to fibre-spaces. Let us consider the direct sum decomposition $\pi_r(S^4) \approx \pi_r(S^7) + \pi_{r-1}(S^3)$ of theorem 2·3 of Chapter V. We recall that this decomposition arises from the decomposition $\pi_r(S^7, S^3) \approx \pi_r(S^7) + \pi_{r-1}(S^3)$ when we apply the isomorphism $\pi_r(S^7, S^3) \approx \pi_r(S^4)$ induced by the fibre-map $S^7 \to S^4$. Now $\pi_{r-1}(S^3)$ is embedded in $\pi_r(S^7, S^3)$ by an isomorphism induced by some correspondence $f \to f'$ between maps $f: S^{r-1} \to S^3$ and extensions of f to $f': E^r, S^{r-1} \to S^7, S^3$. We choose a fixed (but arbitrary) extension $k: E^4, S^3 \to S^7, S^3$ of the identity map $S^3 \to S^3$. Then, if $f'': E^r, S^{r-1} \to E^4, S^3$ is an extension of f, $f' = kf''$ is an extension of f to a map $f': E^r, S^{r-1} \to S^7, S^3$, and it is clear that $f \to f'$ induces the required homomorphism (isomorphism, in fact) of $\pi_{r-1}(S^3)$ into $\pi_r(S^7, S^3)$.

THEOREM 3·1. *The induced embedding of* $\pi_{r-1}(S^3)$ *in* $\pi_r(S^4)$ *is, in fact,* $F: \pi_{r-1}(S^3) \to \pi_r(S^4)$.

† The result due to Adem was communicated to me, as this section was being written, by J.-P. Serre.

We prove this first if $r=4$. Then k represents a generator of $\pi_4(S^7, S^3)$, so that† if $\psi: S^7 \to S^4$ is the fibre-map, the map $\psi k: E^4, S^3 \to S^4, p_0$ generates $\pi_4(S^4)$, and, if S^4 is suitably oriented, ψk represents the positive generator of $\pi_4(S^4)$. Thus the isomorphic embedding of $\pi_3(S^3)$ in $\pi_4(S^4)$ (in this case, of course, an isomorphism on to $\pi_4(S^4)$) is just $F: \pi_3(S^3) \to \pi_4(S^4)$.

It follows that ψk is in the same class as the map

$$\phi_4: E^4, S^3 \to S^4, p_0.$$

We recall from theorem 2·1 that if $f: S^{r-1} \to S^3$ represents $\alpha \in \pi_{r-1}(S^3)$ and if $f'': E^r, S^{r-1} \to E^4, S^3$ is any extension of f, then $\phi_4 f'': E^r, S^{r-1} \to S^4, p_0$ represents $F(\alpha)$. However, we know that the embedding of $\pi_{r-1}(S^3)$ in $\pi_r(S^4)$ is induced by the correspondence $f \to \psi k f''$ and that $\psi k \sim \phi_4: E^4, S^3 \to S^4, p_0$. Thus $\psi k f'' \sim \phi_4 f''$, and $\pi_{r-1}(S^3)$ is embedded in $\pi_r(S^4)$ by F.

THEOREM 3·2. *Let $P: \pi_r(S^7) \to \pi_r(S^4)$ be induced by the projection $\psi: S^7 \to S^4$. Then P and $F: \pi_{r-1}(S^3) \to \pi_r(S^4)$ are isomorphisms and $\pi_r(S^4) = P\pi_r(S^7) + F\pi_{r-1}(S^3)$.*

By precisely analogous arguments we prove

THEOREM 3·3. *Let $P: \pi_r(S^{15}) \to \pi_r(S^8)$ be induced by the projection $\psi: S^{15} \to S^8$. Then P and $F: \pi_{r-1}(S^7) \to \pi_r(S^8)$ are isomorphisms and $\pi_r(S^8) = P\pi_r(S^{15}) + F\pi_{r-1}(S^7)$.*

As a special case we have $\pi_7(S^4) = P\pi_7(S^7) + F\pi_6(S^3)$. The subgroup $P\pi_7(S^7)$ is the cyclic infinite subgroup generated by the class of the Hopf map $S^7 \to S^4$, and $F\pi_6(S^3)$ is the subgroup, isomorphic to $\pi_6(S^3)$, consisting of elements with Hopf invariant 0. Now $F: \pi_7(S^4) \to \pi_8(S^5)$ is on to $\pi_8(S^5)$, $F \mid F\pi_6(S^3)$ is isomorphic and the kernel of F is the cyclic subgroup generated by a particular element with Hopf invariant 2. It follows from this that $\pi_8(S^5)$, and therefore $\pi_{n+3}(S^n)$, $n \geqslant 5$, has a subgroup isomorphic to $\pi_6(S^3)$, and has twice as many elements as $\pi_6(S^3)$; for, if $\alpha \in \pi_7(S^4)$ has Hopf invariant 1, $F(\alpha)$ cannot belong to $F^2\pi_6(S^3)$. A similar analysis applies to $\pi_{15}(S^8)$.

The fact that $F: \pi_{r-1}(S^3) \to \pi_r(S^4)$ and $F: \pi_{r-1}(S^7) \to \pi_r(S^8)$ are always (i.e. for all values of r) isomorphic leads to the following interesting conclusion.

† Changing our notation from Chapter V, we use ψ for the fibre-map, to avoid confusion with $\phi_n: E^n, S^{n-1} \to S^n, p_0$.

THEOREM 3·4. *If* $\alpha \in \pi_r(S^n)$, $\beta_1, \beta_2 \in \pi_n(S^k)$, $k=3$ *or* 7, *then*

$$(\beta_1+\beta_2)\circ\alpha=\beta_1\circ\alpha+\beta_2\circ\alpha.$$

For

$$\begin{aligned}F((\beta_1+\beta_2)\circ\alpha)&=F(\beta_1+\beta_2)\circ F(\alpha)=(F(\beta_1)+F(\beta_2))\circ F(\alpha)\\&=F(\beta_1)\circ F(\alpha)+F(\beta_2)\circ F(\alpha), \quad \text{by theorem 2·3},\\&=F(\beta_1\circ\alpha)+F(\beta_2\circ\alpha)=F((\beta_1\circ\alpha)+(\beta_2\circ\alpha)).\end{aligned}$$

Thus (for all values of k), $(\beta_1+\beta_2)\circ\alpha-\beta_1\circ\alpha-\beta_2\circ\alpha$ lies in the kernel of $F:\pi_r(S^k)\to\pi_{r+1}(S^{k+1})$, and is therefore 0 if $k=3$ or 7.

THEOREM 3·5. *If* $\alpha\in\pi_r(S^n)$, $\beta_1,\beta_2\in\pi_n(S^2)$, $n\geqslant 3$, *then*

$$(\beta_1+\beta_2)\circ\alpha=\beta_1\circ\alpha+\beta_2\circ\alpha.$$

Let $\gamma\in\pi_3(S^2)$ be the class of the Hopf map. Then it follows from theorem 2·1 of Chapter V that $\beta_i=\gamma\circ\beta'_i$, $\beta'_i\in\pi_n(S^3)$, $i=1,2$. Hence

$$\begin{aligned}(\beta_1+\beta_2)\circ\alpha&=(\gamma\circ\beta'_1+\gamma\circ\beta'_2)\circ\alpha=\gamma\circ(\beta'_1+\beta'_2)\circ\alpha\\&=\gamma\circ(\beta'_1\circ\alpha+\beta'_2\circ\alpha), \quad \text{by theorem 3·4},\\&=\gamma\circ\beta'_1\circ\alpha+\gamma\circ\beta'_2\circ\alpha=\beta_1\circ\alpha+\beta_2\circ\alpha.\end{aligned}$$

Theorem 3·5 leaves out of account the case $n=2$. Here the situation is different.

THEOREM 3·6. *If* $\alpha\in\pi_r(S^2)$, *and* i *generates* $\pi_2(S^2)$, *then, for any integer* k,

$$\begin{aligned}(ki)\circ\alpha&=k\alpha, \quad r=2\\&=k^2\alpha, \quad r>2.\end{aligned}$$

The case $r=2$ is trivial. Assume $r>2$, so that $\alpha=\gamma\circ\alpha'$, $\alpha'\in\pi_r(S^3)$. Then $(ki)\circ\alpha=(ki)\circ\gamma\circ\alpha'=k^2\gamma\circ\alpha'$, since the Hopf invariant of $ki\circ\gamma$ is k^2 and $\pi_3(S^2)$ is cyclic infinite. By theorem 3·5, $k^2\gamma\circ\alpha'=k^2(\gamma\circ\alpha')=k^2\alpha$, which proves the theorem.

It may also be noted that it follows almost immediately from the relation $\alpha=\gamma\circ\alpha'$, and the fact that $F(2\gamma)=0$, that

$$2F\pi_r(S^2)=0$$

for all $r\geqslant 3$.

We now apply suspension theory to the consideration of the Stiefel manifolds $V_{n,m}$. We will, in fact, confine attention to $V_{n,2}$ and $V_{n,n-1}$ $(=\Omega_n)$.

We recall from Chapter V that $\pi_3(\Omega_2)=0$, $\pi_3(\Omega_3)=Z_\infty$, $\pi_3(\Omega_4)=Z_\infty+Z_\infty$, $\pi_3(\Omega_5)=\ldots=\pi_3(\Omega_n)=Z_\infty$, and that $\pi_3(\Omega_n)$, $n\geqslant 5$, is generated by the class of a map $f:S^3\to\Omega_4$ which is a cross-section for the fibring of Ω_4 by fibres Ω_3 with base-space S^3. We now prove

THEOREM 3·7. $\pi_4(\Omega_2)=0$, $\pi_4(\Omega_3)=Z_2$, $\pi_4(\Omega_4)=Z_2+Z_2$, $\pi_4(\Omega_5)=Z_2$.

Since Ω_2 is the circle, $\pi_4(\Omega_2)=0$; Ω_3 is real projective 3-space and is covered by S^3, so that, by theorem 2·16, $\pi_4(\Omega_3)=Z_2$; by (3·15) of Chapter V, $\pi_4(\Omega_4)\approx\pi_4(S^3)+\pi_4(S^3)=Z_2+Z_2$. We now calculate $\pi_4(\Omega_5)$.

We note first that in the direct sum decomposition

$$\pi_4(\Omega_4)\approx\pi_4(\Omega_3)+\pi_4(S^3),$$

the isomorphic embedding of $\pi_4(S^3)$ in $\pi_4(\Omega_4)$ is induced by the cross-section $f:S^3\to\Omega_4$. Now consider the diagram

$$\begin{array}{ccccc} & & \pi_3(\Omega_4) & \xrightarrow{i_3} & \pi_3(\Omega_5) \\ & & \downarrow{\scriptstyle\lambda} & & \downarrow{\scriptstyle\mu} \\ \pi_5(\Omega_5,\Omega_4) & \xrightarrow{d} & \pi_4(\Omega_4) & \xrightarrow{i_4} & \pi_4(\Omega_5) \end{array}$$

In the diagram λ and μ are induced by a map $S^4\to S^3$ in the non-zero class of $\pi_4(S^3)$. It follows from theorem 2·3 that λ and μ are homomorphisms, since the map $S^4\to S^3$ is the suspension of a map $S^3\to S^2$. We also need the following facts: (i) λ maps $\pi_3(\Omega_4)$ on to $\pi_4(\Omega_4)$, (ii) i_4 maps $\pi_4(\Omega_4)$ on to $\pi_4(\Omega_5)$, (iii) $\mu i_3=i_4\lambda$, (iv) $i_4^{-1}(0)=d\pi_5(\Omega_5,\Omega_4)$, (v) $\pi_5(\Omega_5,\Omega_4)=Z_2$.

Proof of (i). Since the double-covering $S^3\to\Omega_3$ induces an isomorphism $\pi_n(S^3)\approx\pi_n(\Omega_3)$, it follows that any class in $\pi_n(\Omega_3)$ may be represented by a map $S^n\to S^3\to\Omega_3$. Thus the two classes in $\pi_4(\Omega_3)$ are the zero class and the class containing $S^4\to S^3\to\Omega_3$, where $S^4\to S^3$ is a map in the non-zero class of $\pi_4(S^3)$ and $S^3\to\Omega_3$ is the covering map. We have $\pi_n(\Omega_4)=i\pi_n(\Omega_3)+f^*\pi_n(S^3)$, where i is the injection, and f^* is induced by the cross-section $f:S^3\to\Omega_4$.

What we have proved is that λ maps $i\pi_3(\Omega_3)$ on to $i\pi_4(\Omega_3)$, and the remark immediately before the diagram means that λ maps $f^*\pi_3(S^3)$ on to $f^*\pi_4(S^3)$. This establishes (i). (ii) is a special case ($n=4$, $m=3$) of theorem 3·9 of Chapter V, (iii) is obvious, (iv) follows from the exactness of the homotopy sequence, and (v) follows from the fact that Ω_5 is a fibre-space over S^4, with fibre Ω_4, and the fact that $\pi_5(S^4)=Z_2$.

Since $i_4\lambda$ maps $\pi_3(\Omega_4)$ on to $\pi_4(\Omega_5)$, and $\mu i_3=i_4\lambda$, it follows that μ maps $\pi_3(\Omega_5)$ on to $\pi_4(\Omega_5)$. Since $\pi_3(\Omega_5)$ is cyclic (theorem 3·16 of Chapter V), and μ is a homomorphism, $\pi_4(\Omega_5)$ is cyclic. Now $\pi_4(\Omega_4)=Z_2+Z_2$, so that $\pi_4(\Omega_5)$, as a cyclic homomorphic image, must be 0 or Z_2. If $\pi_4(\Omega_5)$ were zero, it would follow that d maps $\pi_5(\Omega_5, \Omega_4)$ on to $\pi_4(\Omega_4)$, which is impossible, since $\pi_5(\Omega_5, \Omega_4)$ has only two elements. Thus $\pi_4(\Omega_5)=Z_2$. It is generated by the class of a map $S^4\rightarrow S^3\rightarrow\Omega_4$, where $S^4\rightarrow S^3$ is in the non-zero class of $\pi_4(S^3)$ and $S^3\rightarrow\Omega_4$ is the cross-section.

We can now deduce from the exactness of the homotopy sequence of the pair (Ω_5, Ω_4) that $i_5:\pi_5(\Omega_4)\rightarrow\pi_5(\Omega_5)$ is on to $\pi_5(\Omega_5)$. Using the fact that $\pi_{n+2}(S^n)=Z_2$, $n\geqslant 2$, we can deduce that $\pi_5(\Omega_2)=0$, $\pi_5(\Omega_3)=Z_2$, $\pi_5(\Omega_4)=Z_2+Z_2$, $\pi_5(\Omega_5)=Z_2$, but we do not go into details.

It may also be shown that $\pi_4(\Omega_6)$, which is a homomorphic image of $\pi_4(\Omega_5)$, is, in fact, zero, so that $\pi_4(\Omega_n)=0$, $n\geqslant 6$. The proof is difficult and is given by Steenrod, *Topology of Fibre Bundles*, p. 124.

We now turn attention to $\pi_r(V_{n+1,2})$. We will assume n to be even, since, as we have shown in (3·8) of Chapter V, if n is odd, $\pi_r(V_{n+1,2})\approx\pi_r(S^n)+\pi_r(S^{n-1})$ and the problem of computing $\pi_r(V_{n+1,2})$ in this case is reduced to that of calculating homotopy groups of spheres. We prove†

THEOREM 3·8. *If n is even and* $\geqslant 4$, $\pi_{n-1}(V_{n+1,2})=\pi_n(V_{n+1,2})=Z_2$.

Consider the sequence

$$\pi_{n+1}(S^n)\xrightarrow{\bar{d}}\pi_n(S^{n-1})\xrightarrow{i}\pi_n(V_{n+1,2})\xrightarrow{\bar{j}}\pi_n(S^n)\xrightarrow{\bar{d}}\pi_{n-1}(S^{n-1})$$
$$\xrightarrow{i}\pi_{n-1}(V_{n+1,2}),$$

† Of course, $\pi_1(V_{3,2})=Z_2$, but $\pi_2(V_{3,2})=0$, so that the case $n=2$ is properly excluded from the theorem.

arising from the projection of the fibre-space $V_{n+1,2}$ on to S^n with fibre $V_{n,1}=S^{n-1}$. We proved in Chapter V that $\bar{d}\pi_n(S^n)$ is the subgroup of $\pi_{n-1}(S^{n-1})$ generated by $2\epsilon\pi_{n-1}(S^{n-1})$. Since

$$i:\pi_{n-1}(S^{n-1})\rightarrow\pi_{n-1}(V_{n+1,2})$$

is on to $\pi_{n-1}(V_{n+1,2})$ and the kernel of i is generated by 2, it follows that $\pi_{n-1}(V_{n+1,2})=Z_2$. It now follows from this exact sequence (as a special case of theorem 3·9 of Chapter V) that

$$i:\pi_n(S^{n-1})\rightarrow\pi_n(V_{n+1,2})$$

is on to $\pi_n(V_{n+1,2})$. Since $n\geqslant 4$, $\pi_n(S^{n-1})=Z_2$. The conclusion of the theorem will follow if we can show that $\bar{d}\pi_{n+1}(S^n)=0$. This is a ready consequence of

LEMMA 3·9. *Let $t:S^{n-1}\rightarrow S^{n-1}$ represent the generator of $\bar{d}\pi_n(S^n)$ and let t induce the endomorphism $h:\pi_r(S^{n-1})\rightarrow\pi_r(S^{n-1})$. Then $h=\bar{d}F$, where F is the suspension $F:\pi_r(S^{n-1})\rightarrow\pi_{r+1}(S^n)$ and $\bar{d}$ is the homomorphism $\bar{d}:\pi_{r+1}(S^n)\rightarrow\pi_r(S^{n-1})$.*

Let $\alpha=F\beta$, $\alpha\epsilon\pi_{r+1}(S^n)$, $\beta\epsilon\pi_r(S^{n-1})$. Then α may be represented by a map $E^{r+1},S^r\xrightarrow{f}E^n,S^{n-1}\xrightarrow{\phi_n}S^n,p_0$, where $f\mid S^r$ represents β, and ϕ_n has its usual significance. If $\psi:V_{n+1,2}\rightarrow S^n$ is the fibre-mapping, then $\psi u:E^n,S^{n-1}\rightarrow S^n,p_0$ is homotopic to ϕ_n, where $u:E^n,S^{n-1}\rightarrow V_{n+1,2},S^{n-1}$ is the map used in the proof of theorem 3·7 of Chapter V, and is such that $u\mid S^{n-1}=t$. Let ψ induce the isomorphism $\psi^*:\pi_{r+1}(V_{n+1,2},S^{n-1})\approx\pi_{r+1}(S^n)$. Since $\psi uf:E^{r+1},S^r\rightarrow S^n,p_0$ represents $\alpha\epsilon\pi_{r+1}(S^n)$, it follows that $uf:E^{r+1},S^r\rightarrow V_{n+1,2},S^{n-1}$ represents $\psi^{*-1}(\alpha)$. Now

$$\bar{d}:\pi_{r+1}(S^n)\rightarrow\pi_r(S^{n-1})$$

is just $d\psi^{*-1}$, where d is the boundary homomorphism

$$\pi_{r+1}(V_{n+1,2},S^{n-1})\rightarrow\pi_r(S^{n-1}).$$

Thus $\bar{d}(\alpha)$ is represented by $uf\mid S^r$, i.e. by a map $S^r\rightarrow S^{n-1}\rightarrow S^{n-1}$, where $S^r\rightarrow S^{n-1}$ represents β, and $S^{n-1}\rightarrow S^{n-1}$ is precisely the map t. This means that $\bar{d}(\alpha)=h(\beta)$, or $\bar{d}F(\beta)=h(\beta)$. This completes the proof of the lemma.

Now let n be even and $\geqslant 4$. Then $\bar{d}\pi_{n+1}(S^n)=\bar{d}F(\pi_n(S^{n-1}))$, by theorem 2·10, $=h(\pi_n(S^{n-1}))$, by lemma 3·9, and $t:S^{n-1}\rightarrow S^{n-1}$

is a map of degree 2. Since $\pi_n(S^{n-1}) = F\pi_{n-1}(S^{n-2})$, it follows† from theorem 2·3 that

$$h(\pi_n(S^{n-1})) = 2\pi_n(S^{n-1}) = 0,$$

since $\pi_n(S^{n-1}) = Z_2$. This proves the theorem.

If we use the fact that $\pi_{n+2}(S^n) = Z_2$, $n \geqslant 2$, we may prove by the same methods that $\pi_{n+1}(V_{n+1,2})$ is an extension of Z_2 by Z_2 if n is even and $\geqslant 4$. M. Barratt and G. Paechter have recently proved that, in fact, $\pi_{n+1}(V_{n+1,2}) = Z_4$.

THEOREM 3·10. $\pi_4(V_{n+3,n}) = 0$, $n \geqslant 3$.

By (3·5) of Chapter V it is sufficient to show that $\pi_4(V_{6,3}) = 0$. We use the fact that $\pi_4(\Omega_6) = 0$. By (3·12) of Chapter V, $\pi_4(V_{6,4}) = 0$. Now $V_{6,4}$ is a fibre-space over $V_{6,3}$ with fibre $V_{3,1} = S^2$. Thus we have the homotopy sequence

$$\dots \longrightarrow \pi_4(V_{6,4}) \xrightarrow{\bar{\jmath}} \pi_4(V_{6,3}) \xrightarrow{\bar{d}} \pi_3(S^2) \longrightarrow \dots.$$

By (3·6) of Chapter V, $\pi_4(V_{6,3})$ is a homomorphic image of $\pi_4(V_{5,2})$, so that, by theorem 3·8, $\pi_4(V_{6,3})$ is 0 or Z_2. Thus $\bar{d}\pi_4(V_{6,3})$, being a subgroup of a cyclic infinite group, is zero, whence $\bar{\jmath}$ is on to $\pi_4(V_{6,3})$ and $\pi_4(V_{6,3}) = 0$.

4. The generalized Hopf invariant.

Our main objective in this section is to show that the homomorphism

$$H^*: \pi_r(S^n) \to \pi_{r+1}(S^{2n})$$

defined in the first section does generalize the Hopf invariant. We first study the homotopy groups of the union of spheres with a single common point. We saw in § 6 of Chapter IV that if Y, Y' are spaces with a single common point and if we identify Y with $Y \times y'_0$, Y' with $y_0 \times Y'$, $y_0 \in Y$, $y'_0 \in Y'$, then Y, Y' and $Y \cup Y'$ become subspaces of $Y \times Y'$ and we have

$$\pi_n(Y \cup Y') \approx \pi_n(Y) + \pi_n(Y') + \pi_{n+1}(Y \times Y', Y \cup Y'), \quad n > 1. \tag{4·1}$$

Moreover, $\pi_n(Y)$, $\pi_n(Y')$ are embedded in $\pi_n(Y \cup Y')$ by injection and $\pi_{n+1}(Y \times Y', Y \cup Y')$ is embedded by the boundary (isomorphism)

$$d: \pi_{n+1}(Y \times Y', Y \cup Y') \to \pi_n(Y \cup Y');$$

† In general the endomorphism $h: \pi_r(S^{n-1}) \to \pi_r(S^{n-1})$, n even, is given by $h(\alpha) = 2i \circ \alpha$, where i generates $\pi_{n-1}(S^{n-1})$.

and $\pi_n(Y \cup Y')$ is projected on to $\pi_n(Y)$, $\pi_n(Y')$ by the homomorphisms induced by the projections $Y \cup Y' \to Y$, $Y \cup Y' \to Y'$. If we call the projections ρ, ρ' and the injections i, i', then $\pi_n(Y \cup Y')$ is projected on to $\pi_{n+1}(Y \times Y', Y \cup Y')$ by Q, given by $Q(x) = d^{-1}(x - i\rho x - i'\rho' x)$, $x \in \pi_n(Y \cup Y')$.

Now let $Y = S^p$, $Y' = S^q$. Then

$$\pi_n(S^p \cup S^q) \approx \pi_n(S^p) + \pi_n(S^q) + \pi_{n+1}(S^p \times S^q, S^p \cup S^q).$$

Generalizing (C) of § 2, let $\phi_p : E^p \to S^p$ be defined as before, similarly ϕ_q, and define

$$\phi_{p,q} : E^p \times E^q \to S^p \times S^q \quad \text{by} \quad \phi_{p,q}(x, y) = (\phi_p(x), \phi_q(y)).$$

Then $S^p \times S^q = \phi_{p,q}(E^p \times E^q)$, $\phi_{p,q}$ maps the interior of $E^p \times E^q$ homeomorphically on to $S^p \times S^q - S^p \cup S^q$, and $\phi_{p,q}$ maps the boundary of $E^p \times E^q$ into $S^p \cup S^q$. Thus $\phi_{p,q}$ is a characteristic map for the $(p+q)$-cell which is attached to $S^p \cup S^q$ to form $S^p \times S^q$. Let $\phi_{p,q}$ induce

$$g_r : \pi_r(E^{p+q}, \dot{E}^{p+q}) \to \pi_r(S^p \times S^q, S^p \cup S^q).$$

By theorem 2·14, g_r is an isomorphism on to

$$\pi_r(S^p \times S^q, S^p \cup S^q) \quad \text{if} \quad 2 \leqslant r \leqslant p + q + \min(p, q) - 2,$$

and, by using the homotopy boundary isomorphism

$$d' : \pi_r(E^{p+q}, \dot{E}^{p+q}) \approx \pi_{r-1}(S^{p+q-1}),$$

we have $\quad \pi_r(S^p \cup S^q) \approx \pi_r(S^p) + \pi_r(S^q) + \pi_r(S^{p+q-1})$,

if $2 \leqslant r \leqslant p + q + \min(p, q) - 3$. As an immediate consequence, we have

THEOREM 4·2. *$\pi_r(S^p \cup S^q) \approx \pi_r(S^p) + \pi_r(S^q)$ if $2 \leqslant r \leqslant p + q - 2$.*

We now study more closely the manner in which $\pi_r(S^{p+q-1})$ is embedded in $\pi_r(S^p \cup S^q)$. Consider the map

$$\phi_{p,q} \,|\, (E^p \times E^q)^{\cdot} : (E^p \times E^q)^{\cdot} \to S^p \cup S^q.$$

Identifying $(E^p \times E^q)^{\cdot}$ with S^{p+q-1}, it is clear that this map induces a homomorphism $h_r : \pi_r(S^{p+q-1}) \to \pi_r(S^p \cup S^q)$, satisfying $h_r d' = d g_{r+1}$. Since $\pi_{r+1}(S^p \times S^q, S^p \cup S^q)$ is embedded in $\pi_r(S^p \cup S^q)$ by d, and $g_{r+1} d'^{-1} : \pi_r(S^{p+q-1}) \approx \pi_{r+1}(S^p \times S^q, S^p \cup S^q)$, it follows that $\pi_r(S^{p+q-1})$ is embedded in $\pi_r(S^p \cup S^q)$ by the (isomorphism) h_r, and $\pi_r(S^p \cup S^q)$ is projected on to $\pi_r(S^{p+q-1})$

by $d'g_{r+1}^{-1}Q$. The map $\phi_{p,q} \mid (E^p \times E^q)^{\cdot}$ is a special case† of a 'Whitehead product map', which is defined as follows. Let $\alpha, \beta \in \pi_m(Y), \pi_n(Y)$ be given by

$$f: E^m, \dot{E}^m \to Y, y_0, \quad g: E^n, \dot{E}^n \to Y, y_0$$

respectively. Define $h: (E^m \times E^n)^{\cdot} \to Y$ by $h(a,b) = f(a)$, $a \in E^m$, $b \in \dot{E}^n$, $h(a,b) = g(b)$, $a \in \dot{E}^m$, $b \in E^n$. If we choose a fixed point on $\dot{E}^m \times \dot{E}^n$ as base point, h determines an element in $\pi_{m+n-1}(Y)$. This element depends only on α, β and is written $[\alpha, \beta]$; it is called the *Whitehead product* of α and β. If we identify $\pi_p(S^p)$ with its image under injection into $\pi_p(S^p \cup S^q)$ and similarly treat $\pi_q(S^q)$, then $\phi_{p,q} \mid (E^p \times E^q)^{\cdot}$ represents

$$[i^{(p)}, i^{(q)}] \in \pi_{p+q-1}(S^p \cup S^q),$$

where $i^{(p)}$ generates $\pi_p(S^p)$, $i^{(q)}$ generates $\pi_q(S^q)$. Thus, in the special case $r = p + q - 1$, $p \geq 2$, $q \geq 2$, (4·2) reduces‡ to theorem 4 in J. H. C. Whitehead's paper 'On adding relations to homotopy groups' (*Ann. Math.* 42 (1941), 409–28), where he first defined the product $[\alpha, \beta]$.

We now put $p = q = n$. Then

$$\pi_r(S_1^n \cup S_2^n) \approx \pi_r(S_1^n) + \pi_r(S_2^n) + \pi_r(S^{2n-1}), \tag{4·3}$$

if $2 \leq r \leq 3n - 3$. G. W. Whitehead then defined his generalized Hopf invariant as $\bar{Q}\mu$, where $\mu: \pi_r(S^n) \to \pi_r(S_1^n \cup S_2^n)$ was defined in § 1, and $\bar{Q}, = d'g_{r+1}^{-1}Q$, is the projection $\pi_r(S_1^n \cup S_2^n) \to \pi_r(S^{2n-1})$. We will call the homomorphism $\bar{Q}\mu$ by the letter H, so that H is a homomorphism $\pi_r(S^n) \to \pi_r(S^{2n-1})$, defined if $r \leq 3n - 3$.

THEOREM 4·4. *$H^* = FH$, wherever H has been defined.*

Here F is, of course, the suspension and $H^* = \chi Q \mu$, as defined in § 1. Now $FH = F\bar{Q}\mu = Fd'g_{r+1}^{-1}Q\mu$, so it will certainly be sufficient to show that $Fd'g_{r+1}^{-1} = \chi$. Since g_{r+1} is an isomorphism, this is simply a re-expression of (2·2) in the special case

$$(X^*, X) = (S_1^n \times S_2^n, S_1^n \cup S_2^n).$$

COROLLARY 4·5. *$F^{-1}H^*: \pi_r(S^n) \to \pi_r(S^{2n-1})$ is a homomorphism which is defined whenever $F: \pi_r(S^{2n-1}) \to \pi_{r+1}(S^{2n})$ is an isomor-*

† Another special case was the map used in the proof of theorem 1·6 and the statement of theorem 2·15.

‡ We also need the fact that $[\alpha, \beta]$ is bilinear in α, β if $m \geq 2$, $n \geq 2$, proved by Whitehead in the paper to which we have referred.

phism (on to) (e.g. when $r \leqslant 4n-4$, *or* $r=5$, $n=2$, *or* $r=13$, $n=4$) *and coincides with H if* $r \leqslant 3n-3$.

G. W. Whitehead used (4·3) to throw new light on the left-distributive law. Since $\mu(\alpha)=(i_1^{(n)}+i_2^{(n)})\circ\alpha$, $\alpha\in\pi_r(S^n)$, where $i_\lambda^{(n)}$ generates $\pi_n(S_\lambda^n)$, $\lambda=1,2$, it follows that,† if $r\leqslant 3n-3$,

$$\mu\alpha=(i_1^{(n)}+i_2^{(n)})\circ\alpha=i_1^{(n)}\circ\alpha+i_2^{(n)}\circ\alpha+[i_1^{(n)},i_2^{(n)}]\circ H(\alpha). \quad (4·6)$$

Now let $\beta_1,\beta_2\in\pi_n(X)$, let $f_\lambda: S_\lambda^n\to X$ represent β_λ, $\lambda=1,2$, and let β be the homotopy class‡ of the map $f: S_1^n\cup S_2^n\to X$ given by $f\mid S_\lambda^n=f_\lambda$. It is then clear that $\beta\circ(i_1^{(n)}+i_2^{(n)})=\beta_1+\beta_2$; this follows immediately from the definition of addition in homotopy groups. It is also straightforward to verify that $\beta\circ[i_1^{(n)},i_2^{(n)}]=[\beta_1,\beta_2]$; the representing maps will be the same. Finally, we point out that $\beta\circ i_1^{(n)}=\beta_1$, $\beta\circ i_2^{(n)}=\beta_2$. From (4·6) we have

$$\beta\circ(i_1^{(n)}+i_2^{(n)})\circ\alpha=\beta\circ i_1^{(n)}\circ\alpha+\beta\circ i_2^{(n)}\circ\alpha+\beta\circ[i_1^{(n)},i_2^{(n)}]\circ H(\alpha),$$

so that we have proved

THEOREM 4·7. *If* $\alpha\in\pi_r(S^n)$, $\beta_1,\beta_2\in\pi_n(X)$, $r\leqslant 3n-3$, *then*

$$(\beta_1+\beta_2)\circ\alpha=\beta_1\circ\alpha+\beta_2\circ\alpha+[\beta_1,\beta_2]\circ H(\alpha).$$

Since $H^*=FH:\pi_{2n-1}(S^n)\to\pi_{2n}(S^{2n})$, it is obviously sufficient to show that H generalizes the Hopf invariant, and the same will then follow for H^*. The sense in which H generalizes the Hopf invariant is indicated in the following theorem:

THEOREM 4·8. *If* $\alpha\in\pi_{2n-1}(S^n)$, *then* $H(\alpha)=\pm\gamma(\alpha)\,i^{(2n-1)}$, *where* $\gamma(\alpha)$ *is the Hopf invariant of* α *and* $i^{(2n-1)}$ *generates* $\pi_{2n-1}(S^{2n-1})$.

If n is odd, then $\gamma(\alpha)=0$, and, by theorem 2·15, α is a suspension element. Thus, by theorem 2·9,

$$H^*(\alpha)=0 \quad\text{and}\quad H(\alpha)=F^{-1}H^*(\alpha)=0.$$

Let n be even. Put $X=S^n$, $\beta_1=\beta_2=i^{(n)}$, $r=2n-1$, in theorem 4·7. Then

$$\begin{aligned}2i^{(n)}\circ\alpha&=i^{(n)}\circ\alpha+i^{(n)}\circ\alpha+[i^{(n)},i^{(n)}]\circ H(\alpha)\\&=2\alpha+[i^{(n)},i^{(n)}]\circ H(\alpha).\end{aligned}$$

† It is obvious that, under the projection $S_1^n\cup S_2^n\to S_\lambda^n$, $(i_1^{(n)}+i_2^{(n)})\circ\alpha$ projects on to $i_\lambda^{(n)}\circ\alpha$, $\lambda=1,2$.

‡ Relative to the common point of S_1^n, S_2^n as base-point.

We now take the Hopf invariant of each side and we let $H(\alpha)$ stand now for the characteristic degree of the class

$$H(\alpha) \in \pi_{2n-1}(S^{2n-1}).$$

We will prove that $H(\alpha)$, in this sense, equals $\pm\gamma(\alpha)$.

By lemmas 1·2 and 1·1, $\gamma(2i^{(n)} \circ \alpha) = 4\gamma(\alpha)$, $\gamma(2\alpha) = 2\gamma(\alpha)$, and $\gamma([i^{(n)}, i^{(n)}] \circ H(\alpha)) = H(\alpha) \times \gamma([i^{(n)}, i^{(n)}])$. Now $[i^{(n)}, i^{(n)}]$ is precisely the element which was shown in theorem 1·6 to have Hopf invariant ± 2. Thus $4\gamma(\alpha) = 2\gamma(\alpha) \pm 2H(\alpha)$ or $2\gamma(\alpha) = \pm 2H(\alpha)$. Since $\gamma(\alpha), H(\alpha)$ are integers, we may divide by 2 and get the desired result.

We close this section by giving just one relation between H and properties of homotopy groups of spheres already used. However, we emphasize that the motivation for introducing H and H^* was to obtain non-zero elements of homotopy groups of spheres.

We recall that $\pi_r(S^n) \approx \pi_r(S^{2n-1}) + \pi_{r-1}(S^{n-1})$, $r \geqslant 2$, $n = 2, 4$, or 8. Then there is a projection $H_1 : \pi_r(S^n) \to \pi_r(S^{2n-1})$.

THEOREM 4·9. *If* $r \leqslant 3n-3$, *then* $H_1 = H$.

Let β be the class of the Hopf map $S^{2n-1} \to S^n$. Then, if $\alpha \in \pi_r(S^n)$, $\alpha = \beta \circ \gamma + \delta$, where $\gamma \in \pi_r(S^{2n-1})$, $\delta \in F\pi_{r-1}(S^{n-1})$, and $\gamma = H_1(\alpha)$. Then $H(\delta) = 0$ and $H(\alpha) = H(\beta \circ \gamma)$.

Now

$$\begin{aligned}
H(\beta \circ \gamma) &= \bar{Q}\mu(\beta \circ \gamma) \\
&= \bar{Q}((i_1^{(n)} + i_2^{(n)}) \circ \beta \circ \gamma) \\
&= \bar{Q}((i_1^{(n)} \circ \beta + i_2^{(n)} \circ \beta + [i_1^{(n)}, i_2^{(n)}]) \circ \gamma), \text{ since } H(\beta) \text{ is } i^{(2n-1)}, \\
&= \bar{Q}(i_1^{(n)} \circ \beta \circ \gamma + i_2^{(n)} \circ \beta \circ \gamma + [i_1^{(n)}, i_2^{(n)}] \circ \gamma), \text{ since } \gamma \text{ is a suspension element}, \\
&= \bar{Q}h_r(\gamma), \text{ where } h_r(\gamma) = [i_1^{(n)}, i_2^{(n)}] \circ \gamma, \\
&= \gamma, \text{ since } h_r \text{ injects } \pi_r(S^{2n-1}) \text{ into } \pi_r(S_1^n \cup S_2^n) \text{ and } \bar{Q} \text{ projects } \pi_r(S_1^n \cup S_2^n) \text{ back on to } \pi_r(S^{2n-1}), \\
&= H_1(\alpha).
\end{aligned}$$

In fact, theorem 4·9 may be further generalized to the theorem that, if $H_1(\alpha)$ is a suspension element, then $H^*(\alpha) = FH_1(\alpha)$, but we will not prove that here. We have recently shown that the relation $H^* = FH_1$ does not necessarily hold if $H_1(\alpha)$ is not a suspension element.

CHAPTER VII

WHITEHEAD CELL-COMPLEXES†

1. Definition of a cell-complex, and the basic properties of CW-complexes.

For convenience, since simplicial subdivision is a tedious operation in many cases, and for greater generality, it is advisable to extend our notion of a simplicial complex to the more general notion of a *cell-complex*. This extension is due to J. H. C. Whitehead, who defined a cell-complex‡ as follows.

A cell-complex, K, is a Hausdorff space which is the union of disjoint (open) cells, e^n. The closure, $\bar{e}^n$, of the cell e^n, is the image of an n-element E^n under a map $f: E^n, S^{n-1} \to K, K^{n-1}$ such that $f \mid E^n - S^{n-1}$ is a homeomorphism on to e^n, where K^{n-1} is the point-set union of the cells whose dimension does not exceed $(n-1)$. Thus, in the terminology of Chapter VI, e^n is attached to K^{n-1} by the map $f \mid S^{n-1}$ and f is a characteristic map for e^n. It should be noted that this definition is certainly consistent with the topology of K. For, since E^n is compact and K is Hausdorff, $\bar{e}^n$ is certainly a closed set. Moreover, there can be no closed set F satisfying $e^n \subset F \subset \bar{e}^n$, since there is no closed set $f^{-1}(F)$ satisfying $E^n - S^{n-1} \subset f^{-1}(F) \subset E^n$, the inclusions being, of course, strict inclusions.

A *subcomplex*, L, of K is the union of certain cells of K, such that if $e^n \subset L$, then $\bar{e}^n \subset L$. Thus, if K^r denotes the union of cells of dimension $\leqslant r$, K^r is a subcomplex of K. It should be noted that $\bar{e}^n$ need not be a subcomplex of K, and that a subcomplex of K need not be a closed subset of K. As an example of the latter 'pathology', any Hausdorff space may be considered as a union of 0-cells and then any subset is a subcomplex. To avoid this peculiarity and also in order to be able to carry over many of the

† The material of this chapter is based very largely on J. H. C. Whitehead, 'Combinatorial homotopy, I', *Bull. Amer. Math. Soc.* 55, 3 (1949), 213–45, hereafter referred to as CHI.

‡ Following Whitehead, we do not distinguish between the complex and the underlying space. Thus we dispense with the symbol $|K|$ of Chapter III.

methods and results of the homotopy theory of finite simplicial complexes, Whitehead introduced the following restrictions on the cell-complexes which he studied.

A cell-complex is *closure finite* if, for each e^n, $\bar{e}^n$ is contained in a finite subcomplex. A cell-complex has *the weak topology* if a subset X is closed provided $X \cap \bar{e}^n$ is closed for each cell e^n of the complex. It should be noted that, if K is closure-finite, then it has the weak topology if X is closed provided $X \cap L$ is closed for each finite subcomplex, L, of K. If K is closure-finite and has the weak topology, it is called a *CW-complex*. The *dimension* of K is the least upper bound (perhaps ∞) of the dimensions of its cells. For convenience we will also assume K connected.

The following two examples show the advantage of cell-complexes over simplicial complexes:

(A) $S^n = e^0 \cup e^n$. That is to say, the n-sphere may be decomposed into two cells, a 0-cell e^0 and an n-cell e^n; $\bar{e}^n, = S^n$, is the image of a map $f: E^n, \dot{E}^n \to S^n, e^0$ such that $f \mid E^n - \dot{E}^n$ is a homeomorphism on to $S^n - e^0$.

(B) $\sigma^n \times I = \sigma^n \times 0 \cup \sigma^n \times (0,1) \cup \sigma^n \times 1$. That is to say, if σ^n is an open n-simplex, which is an n-cell, $\sigma^n \times I$ may be decomposed into the two n-cells, $\sigma^n \times 0$ and $\sigma^n \times 1$, and the $(n+1)$-cell $\sigma^n \times (0,1)$. Since $\bar{\sigma}^n = \sigma^n \cup S^{n-1} = \sigma^n \cup e^0 \cup e^{n-1}$, this also gives a cell-decomposition of $\bar{\sigma}^n \times I$ into two 0-cells, $e^0 \times 0$, $e^0 \times 1$; a 1-cell, $e^0 \times (0,1)$; two $(n-1)$-cells, $e^{n-1} \times 0$, $e^{n-1} \times 1$; three n-cells, $e^{n-1} \times (0,1)$, $\sigma^n \times 0$, $\sigma^n \times 1$; and an $(n+1)$-cell, $\sigma^n \times (0,1)$.

In CHI, J. H. C. Whitehead lists a number of properties of CW-complexes. Among the most important for our purposes are the following. The proofs are given in CHI.

THEOREM 1·1. *If $X \subset K$ is compact, then X is contained in a finite subcomplex of K.*

THEOREM 1·2. *If K, L are CW-complexes, and if, moreover, L is locally finite* (i.e. *each point of L is an inner point of a finite subcomplex*), *then $K \times L$ is a CW-complex.*

The cells of $K \times L$ are $e \times e'$, where e is a cell of K, e' a cell of L. We need this theorem principally† when $L = I$. When, as in this

† An example due to C. H. Dowker shows that the theorem would be false if no further restriction were placed on L, beyond the fact it was a CW-complex.

case, L is a finite complex, the proof given in CHI may, of course, be greatly simplified.

THEOREM 1·3. (i) *A transformation* $f: X \to Y$ *of a closed subset* X, *of* K, *into* Y *is continuous provided* $f \mid X \cap \bar{e}$ *is continuous for each cell* e *of* K. (ii) *A set of transformations* $f_t: X \to Y$ *of a closed subset* X, *of* K, *into* Y *is a homotopy provided* $f_t \mid X \cap \bar{e}$ *is a homotopy for each cell* e *of* K.

(i) is a fairly easy consequence of the fact that K has the weak topology. (ii) follows from theorem 1·2, with $L = I$, and (i).

THEOREM 1·4. (The homotopy extension theorem† for CW-complexes.) *Let* $f_0: K \to Y$ *and let* $g_t: L \to Y$ *be a homotopy of* $g_0 = f_0 \mid L$, *where* L *is a subcomplex of* K. *Then there is a homotopy* $f_t: K \to Y$ *with* $f_t \mid L = g_t$.

We give the proof of this since it is, in a sense, representative of arguments about CW-complexes, and because the proof in CHI requires reference to an earlier paper of J. H. C. Whitehead. As pointed out in CHI, we require the following lemma:

LEMMA 1·5. *Let* $K = K^{n-1} \cup e^n$, $L = K^{n-1}$ *in theorem* 1·4. *Then the theorem holds.*

Let $G': K \times 0 \cup K^{n-1} \times I \to Y$ be given by

$$G'(p, 0) = f_0(p), \quad p \in K, \quad G'(q, t) = g_t(q), \quad q \in K^{n-1}.$$

Let‡ $h: E^n, \dot{E}^n \to K, K^{n-1}$ be a characteristic map for e^n, so that $h \mid E^n - \dot{E}^n$ is a homeomorphism on to e^n, and define

$$H: E^n \times I \to K \times I \quad \text{by} \quad H(x, t) = (h(x), t), \quad x \in E^n.$$

Let $T': E^n \times 0 \cup \dot{E}^n \times I \to Y$ be given by $T' = G'H$. By lemma 3·1 of Chapter II, T' may be extended to $T: E^n \times I \to Y$. Define $G: K \times I \to Y$ by $G \mid K^{n-1} \times I = G'$, $G \mid \bar{e}^n \times I = TH^{-1}$. Then G is single-valued, since, if $p \in \bar{e}^n - e^n$,

$$TH^{-1}(p, t) = G'HH^{-1}(p, t) = G'(p, t).$$

It is also clear that G is an extension of G', so that it only remains to show that G is continuous. It is sufficient to show that $G \mid \bar{e}^n \times I$ is continuous. Let us write $F = TH^{-1}: \bar{e}^n \times I \to Y$. If Z

† Cf. theorem 4·9 of Chapter II.

‡ We use the symbol $\dot{E}^n$ rather than S^{n-1} for the boundary of E^n to stress the fact that E^n need not be a Euclidean n-element.

is any subset of Y, it follows from the fact that F is single-valued that $F^{-1}(Z)=H(T^{-1}(Z))$. If Z is closed, $T^{-1}(Z)$ is closed and therefore compact since $E^n \times I$ is compact. Since H is continuous and $K \times I$ is a Hausdorff space, it follows that $H(T^{-1}(Z))$ is closed, so that F is continuous. This proves the lemma.

To prove the theorem, we let $K_r = L \cup K^r$, where K^{-1} is the empty set. Define $f_t^{-1}=g_t: L \to Y$ and suppose we have extended g_t to $f_t^{n-1}: K_{n-1} \to Y$ with $f_0^{n-1}=f_0 \mid K_{n-1}$. By the lemma, we can extend f_t^{n-1} throughout each $e^n \in K-L$, and by theorem 1·3 (ii), the resulting homotopy $f_t^n: K_n \to Y$ will be continuous. Thus we get a sequence of homotopies

$$f_t^n: K_n \to Y, \quad f_0^n = f_0 \mid K_n, \quad f_t^n \mid K_{n-1} = f_t^{n-1},$$

and the required homotopy $f_t: K \to Y$ is given by $f_t \mid K_n = f_t^n$.

We may prove, in a very similar way, that the space Y has vanishing homotopy groups up to dimension n, if and only if every map of an n-dimensional CW-complex into Y is homotopic to a constant map. Combining this with theorem 1·4, we have the following theorem:

THEOREM 1·6. $\pi_r(Y)=0$, $r=1,\ldots,n$, *if and only if every mapping of a CW-complex K into Y is homotopic to a map f such that f maps K^n to a single point.*

This theorem may be relativized as follows:

THEOREM 1·7. $\pi_r(Y, Y_0)=0$, $r=1,\ldots,n$, *if and only if every mapping of a CW-complex K into Y is homotopic to a map f such that f maps K^n into Y_0.*

If L is a subcomplex of K, and if a mapping $K, L \to Y, Y_0$ is given, the construction underlying theorem 1·7 clearly allows us to choose the homotopy so that the image of L remains pointwise fixed throughout the homotopy. We also note that, in theorems 1·6 and 1·7, n may be allowed to take the value ∞.

The map $f: K \to P$, where P is also a cell-complex, is called *cellular* if $f(K^n) \subset P^n$ for each n.

THEOREM 1·8. (Cellular approximation theorem.) *Let*

$$f_0: K \to P$$

be a map of K into the CW-complex P such that $f_0 \mid L$ is cellular, where L is a subcomplex of K. Then $f_0 \sim f_1 : K \to P$, rel L, *where f_1 is cellular.*

We prove† this theorem if $K = \sigma^n$, $L = \dot{\sigma}^n$. The proof in the general case then follows from property (K), CHI, p. 228.

By theorem 1·1, $f_0(\sigma^n)$, being compact, is contained in a finite subcomplex, Q, of P. Thus f_0 is a map $f_0 : \sigma^n, \dot{\sigma}^n \to Q, Q^{n-1}$. Let Q be m-dimensional, $m > n$, let e^m be an m-cell of Q and let $q_0 \in e^m$. Let E be a triangulation of σ^n, so fine that if σ is a simplex of E intersecting $f_0^{-1}(q_0)$, then $f_0(\bar{\sigma}) \subset e^m$. Let A be the subcomplex of E consisting of those closed simplexes which intersect $f_0^{-1}(q_0)$ and let $B = \overline{E - A}$. Then $f_0(B) \subset Q - q_0$, so that

$$f_0(A \cap B) \subset e^m - q_0.$$

Since A is at most n-dimensional and since $\pi_k(e^m - q_0) = 0$, $k = 1, \ldots, m-2 \geqslant n-1$, it follows that $f_0 \mid A \cap B$ may be extended, step-by-step through the simplexes of $A - (A \cap B)$, to a map $f'_0 : A \to e^m - q_0$. Since f_0, f'_0 are both maps $A \to e^m$, they are homotopic, and we may extend the homotopy throughout E by defining it as f_0 on B. Thus $f_0 \sim f'_0 : \sigma^n \to Q$, rel $\dot{\sigma}^n$, and

$$f'_0 \sigma^n \subset Q - q_0.$$

Next, we pull the image of σ^n right off e^m. Precisely, let $g : E^m, \dot{E}^m \to Q, Q^{m-1}$ be a characteristic map for e^m, and let $g(p_0) = q_0$. Let $\rho_t : E^m - p_0 \to E^m - p_0$ be the radial projection of $E^m - p_0$ on to $\dot{E}^m$, and define $\theta_t : Q - q_0 \to Q - q_0$ by $\theta_t \mid Q - e^m = 1$, $\theta_t \mid \bar{e}^m - q_0 = g\rho_t g^{-1}$. θ_t is single-valued since $\bar{e}^m - e^m = g(\dot{E}^m)$ and $g\rho_t g^{-1}(g(p)) = g\rho_t(p) = g(p)$, $p \in \dot{E}^m$. As in the proof of lemma 1·5, it follows that θ_t is continuous, and, of course, $\theta_0 = 1$. Thus $\theta_t f'_0$ is a homotopy of f'_0 and $\theta_1 f'_0 \sigma^n \subset Q - e^m$. Moreover, since $f'_0 \dot{\sigma}^n \subset Q - e^m$, $\theta_1 f'_0 \sim f'_0$, rel $\dot{\sigma}^n$.

In this way, the image of σ^n may be pulled off each m-cell of Q. If $m - 1 > n$, we then repeat the process, until eventually we arrive at a map f_1 such that $f_1 \sim f_0 : \sigma^n \to Q$, rel $\dot{\sigma}^n$ and $f_1 \sigma^n \subset Q^n$.

Let $f_0, f_1 : K \to P$ be cellular. Then we say that a homotopy $f_t : K \to P$ is *cellular* if $f_t(K^n) \subset P^{n+1}$ for all n, t.

† We actually use this theorem often when L is empty.

THEOREM 1·9. *If $f_0 \sim f_1 : K \to P$ and if the homotopy restricted to L, a subcomplex of K, is cellular, then the homotopy may be replaced by a cellular homotopy, extending the cellular homotopy on L.*

This follows immediately from theorem 1·8 with K, L, of that theorem, replaced by $K \times I, K \times 0 \cup K \times 1 \cup L \times I$.

Let K be a locally connected and locally simply-connected complex, and $\tilde{K}$ a covering space of K. This covering space may be made into a cell-complex, as shown in CHI. In fact, if $p : \tilde{K} \to K$ is the covering map, then each cell of $\tilde{K}$ is mapped homeomorphically by p on to a cell of K.

THEOREM 1·10. *If K is a CW-complex, so is any $\tilde{K}$ covering K.*

2. The n-type of a complex and the Massey homology spectrum. Let f, g be maps of X into Y. Then we say that f and g are *n-homotopic* ($f \underset{n}{\simeq} g$) if, for every map ϕ of an arbitrary CW-complex P, of dimension $\leqslant n$, into X, $f\phi \sim g\phi$. We may thus also define *n-homotopy type*, saying that X and Y are of the same n-homotopy type if there exist maps $f : X \to Y, g : Y \to X$ such that $fg \underset{n}{\simeq} 1, gf \underset{n}{\simeq} 1$.

Now let K, L be CW-complexes. We then say that K and L are of the same *n-type* if and only if K^n and L^n are of the same $(n-1)$-homotopy type.

THEOREM 2·1. *Let f_0, f_1 be maps of a CW-complex, K, into a space X. Then $f_0 \underset{n}{\simeq} f_1$ if and only if $f_0 \mid K^n \sim f_1 \mid K^n$.*

The necessity of the condition is clear. Now let ϕ be a map of an arbitrary CW-complex P^n into K. By theorem 1·8,

$$\phi \sim \phi' : P^n \to K$$

with $\phi' P^n \subset K^n$. Thus, if

$$f_0 \mid K^n \sim f_1 \mid K^n, \quad f_0\phi \sim f_0\phi' \sim f_1\phi' \sim f_1\phi.$$

THEOREM 2·2. *If K, L are of the same n-type, then they are of the same m-type, $m \leqslant n$.*

Let $f : K^n \to L^n, g : L^n \to K^n$ be such that $fg \underset{n-1}{\simeq} 1$, $gf \underset{n-1}{\simeq} 1$. We may, by theorem 1·8, assume that f, g are cellular, and we may further assume, by theorem 1·9, that the homotopies

$$fg \mid L^{n-1} \sim 1 : L^{n-1} \to L^n, \quad gf \mid K^{n-1} \sim 1 : K^{n-1} \to K^n$$

are cellular. It is then clear that $f', = f \mid K^m$, and $g', = g \mid L^m$, are such that $f'g' \underset{m-1}{\simeq} 1$, $g'f' \underset{m-1}{\simeq} 1$.

We call such a map f an *n-equivalence* and g its *n-inverse*. Note that we can allow n to take the conventional value ∞, in which case n-homotopy type and n-type coincide with homotopy type. Theorem 2·2 then shows that n-type is an invariant of homotopy type.

It may be shown that connected CW-complexes are of the same 2-type if and only if they have isomorphic fundamental groups. In a recent paper (*Proc. Nat. Acad. Sci., Wash.*, 36 (1950, 1), S. Maclane and J. H. C. Whitehead give an algebraic characterization of 3-type. More recently still,† M. M. Postnikov has given a complete set of invariants of n-type (including the case $n = \infty$).

Still following J. H. C. Whitehead in CHI, we define‡ a *J_m-complex* as one in which the injection $i: \pi_n(K^{n-1}) \to \pi_n(K^n)$ maps $\pi_n(K^{n-1})$ to 0, $n = 2, \ldots, m$.

THEOREM 2·3. *The property of being a J_m-complex is an invariant of m-type.*

In fact, let K be a J_m-complex and let $f: K^m \to L^m$, $g: L^m \to K^m$ be such that $fg \underset{m-1}{\simeq} 1$. We may assume f, g cellular, so that they induce homomorphisms

$$\phi_n: i\pi_n(K^{n-1}) \to i\pi_n(L^{n-1}), \quad \psi_n: i\pi_n(L^{n-1}) \to i\pi_n(K^{n-1}), \ n \leqslant m.$$

We may also take the homotopy $fg \mid L^{m-1} \sim 1: L^{m-1} \to L^m$ as cellular, so that $fg \mid L^{n-1}: L^{n-1} \to L^n$ induces an automorphism of $i\pi_n(L^{n-1})$. Thus $\phi_n \psi_n$ is an automorphism, so that ϕ_n maps $i\pi_n(K^{n-1})$ on $i\pi_n(L^{n-1})$. Thus, if K is a J_m-complex, so is L.

THEOREM 2·4. *If K is a simply-connected J_m-complex, and if $\omega_n: \pi_n(K) \to H_n(K)$ is the natural homomorphism of homotopy groups into homology groups, then ω_n is an isomorphism (on to) if $n \leqslant m$, and ω_{m+1} is on to.*

† See *Doklady Akad. Nauk SSSR*, 76, 3; 76, 6; 79, 4 (1951).

‡ J. H. C. Whitehead's definition is actually different from this, but, as he points out explicitly, equivalent to it.

We first make the following observation, whose importance is by no means confined to this theorem. Since, clearly,

$$H_r(K^n, K^{n-1}) \quad \text{and} \quad \pi_r(K^n, K^{n-1})$$

vanish if $r \leqslant n-1$, and since $\pi_1(K)=0$, it follows from theorem 3·4 of Chapter III, extended to cell-complexes, that there is a natural isomorphism of $\pi_n(K^n, K^{n-1})$† on to $H_n(K^n, K^{n-1})$. Since the latter is the nth chain group we may identify

$$\pi_n(K^n, K^{n-1})$$

with the group of n-dimensional chains of K. In fact, we identify e^n, as a generator of the nth chain group, with the homotopy class of its characteristic map $f: E^n, S^{n-1} \to K^n, K^{n-1}$. Let $d_n: \pi_n(K^n, K^{n-1}) \to \pi_{n-1}(K^{n-1})$ be the homotopy boundary homomorphism and let $j_{n-1}: \pi_{n-1}(K^{n-1}) \to \pi_{n-1}(K^{n-1}, K^{n-2})$ be the injection. Then $\delta_n = j_{n-1} d_n: \pi_n(K^n, K^{n-1}) \to \pi_{n-1}(K^{n-1}, K^{n-2})$ is the homology boundary and $j\pi_{n-1}(K^{n-1})$ is the group of spherical $(n-1)$-cycles.

We now characterize the homomorphism $\omega_n: \pi_n(K) \to H_n(K)$ for a simply-connected complex K. It follows immediately from theorem 1·8 that the injection $i_n: \pi_n(K^n) \to \pi_n(K)$ is on to $\pi_n(K)$ and that the injection $\pi_n(K^{n+1}) \to \pi_n(K)$ is an isomorphism. Given any element $\alpha \in \pi_n(K)$, we may recognize $\omega_n(\alpha)$ as the homology class of $j_n(i_n^{-1}(\alpha))$. (Note that, hitherto, we have only defined ω_n for simplicial complexes.) Then ω_n is single-valued, since, if $\beta \in i_n^{-1}(0)$, then, by exactness,

$$\beta = d_{n+1}\gamma, \quad \gamma \in \pi_{n+1}(K^{n+1}, K^n) \quad \text{and} \quad j_n\beta = \delta_{n+1}\gamma.$$

Thus a different choice of element in $i_n^{-1}(\alpha)$ alters $j_n(i_n^{-1}(\alpha))$ within its homology class. Also $j_n\eta$ is a cycle, $\eta \in \pi_n(K^n)$, because $\delta_n j_n = j_{n-1}(d_n j_n) = 0$, by exactness.

We now prove the theorem, so that K is a simply-connected J_m-complex. Since the injection

$$\pi_n(K^{n-1}) \to \pi_n(K^n)$$

is on to 0, $n \leqslant m$, it follows by exactness that

$$j_n: \pi_n(K^n) \to \pi_n(K^n, K^{n-1})$$

† We make $\pi_n(K^n, K^{n-1})$ abelian if $n=1$ or 2.

is an isomorphism (into) if $n \leqslant m$. Thus, if $n \leqslant m+1$,

$$\delta_n^{-1}(0) = d_n^{-1}\ (j_{n-1}^{-1}(0)) = d_n^{-1}(0) = j_n \pi_n(K^n).$$

Thus j_n maps $\pi_n(K^n)$ on to the group of n-cycles, whence ω_n is on to $H_n(K)$, $n \leqslant m+1$.

Now let $\omega_n \alpha = 0$, $n \leqslant m$. Then there exists $\beta \in \pi_n(K^n)$ such that $i_n \beta = \alpha$ and $j_n \beta = \delta_{n+1}\gamma = j_n d_{n+1}\gamma$, $\gamma \in \pi_{n+1}(K^{n+1}, K^n)$. Since j_n is isomorphic, $n \leqslant m$, we have $\beta = d_{n+1}\gamma$, and $\alpha = i_n \beta = i_n d_{n+1}\gamma = 0$, by exactness. This proves the theorem.

Now suppose $\pi_r(K) = 0$, $r = 1, \ldots, m-1$. Then it is not difficult† to see that, if e^0 is a single point, the map $f: K^m \to e^0$ is an m-equivalence. Since e^0 is certainly a J_m-complex, so is K^m and hence K. Thus we get

THEOREM 2·5. (Hurewicz's theorem for cell-complexes.) *If K is a CW-complex such that $\pi_r(K) = 0$, $r = 1, \ldots, n-1$, then $\omega_n : \pi_n(K) \approx H_n(K)$ and ω_{n+1} maps $\pi_{n+1}(K)$ on to $H_{n+1}(K)$.*

It is interesting to note that theorem 2·5 gives the additional information about ω_{n+1}. It may also be observed that theorem 2·4 is strictly more general than theorem 2·5. For if we take K as $S^3 \cup e^4$, where the characteristic map for e^4 is of degree 3 over S^3, then it may be shown that K is a J_5-complex, but $\pi_3(K) \neq 0$.

Theorem 2·5 may be relativized as follows:

THEOREM 2·6. *If K is a CW-complex, L a subcomplex such that $\pi_1(L) = 0$, $\pi_r(K, L) = 0$, $r = 1, \ldots, n-1$, then the natural homomorphism $\omega_n : \pi_n(K, L) \to H_n(K, L)$ is an isomorphism (on to) and ω_{n+1} maps $\pi_{n+1}(K, L)$ on to $H_{n+1}(K, L)$.*

Now consider the system of groups $\{\pi_r(K^n, K^{n-1}); \pi_r(K^n)\}$ of a CW-complex K and the related homomorphisms

$$d_r^n : \pi_r(K^n, K^{n-1}) \to \pi_{r-1}(K^{n-1}),$$

$$j_r^n : \pi_r(K^n) \to \pi_r(K^n, K^{n-1}),$$

$$i_r^n : \pi_r(K^n) \to \pi_r(K^{n+1}),$$

$$\delta_r^n = j_{r-1}^{n-1} d_r^n : \pi_r(K^n, K^{n-1}) \to \pi_{r-1}(K^{n-1}, K^{n-2}).$$

† See, for example, theorem 3·6 of this chapter, though a direct elementary proof is also available, based on theorem 1·6.

Now, it follows from the exactness of the homotopy sequence of the pair (K^n, K^{n-1}) that $d_r^n j_r^n = 0$, so that

$$\delta_r^n \delta_{r+1}^{n+1} = j_{r-1}^{n-1} d_r^n j_r^n d_{r+1}^{n+1} = 0.$$

Thus we have a homology theory based on the 'chain' groups $\pi_r(K^n, K^{n-1})$ for each fixed value of $(r-n)$. Of course the groups $\pi_r(K^n, K^{n-1})$ are all zero if $r-n<0$; if $r=n$ and K is simply-connected we get the usual homology theory of K. Let H_r^n be the homology group 'based' on the chain group $\pi_r(K^n, K^{n-1})$ and let $\Gamma_r^n = i_r^{n-1}\pi_r(K^{n-1}) \subset \pi_r(K^n)$. We note that, if $r-n<0$, $\pi_r(K^n) = \pi_r(K)$, and, if $r=n$, $\Gamma_r^{n+1} = \pi_r(K)$.

It is convenient at this stage to draw the diagram representing the groups $\pi_r(K^n, K^{n-1})$, $\pi_r(K^n)$, and the related homomorphisms:

$$\begin{array}{ccccccccc}
 & & \ldots \to \pi_r(K^{n-1}) & \to & \pi_r(K^{n-1}, K^{n-2}) & \to & \pi_{r-1}(K^{n-2}) & \to \ldots & \\
 & & \downarrow{\scriptstyle i_r^{n-1}} & & & & \downarrow{\scriptstyle i_{r-1}^{n-2}} & & \\
(K^{n+1}, K^n) & \xrightarrow{d_{r+1}^{n+1}} & \pi_r(K^n) & \xrightarrow{j_r^n} & \pi_r(K^n, K^{n-1}) & \xrightarrow{d_r^n} & \pi_{r-1}(K^{n-1}) & \xrightarrow{j_{r-1}^{n-1}} & \pi_{r-1}(K^{n-1}, K^{n-2}) \\
 & & \downarrow{\scriptstyle i_r^n} & & & & \downarrow{\scriptstyle i_{r-1}^{n-1}} & & \\
 & & \ldots \to \pi_r(K^{n+1}) & \xrightarrow{j_r^{n+1}} & \pi_r(K^{n+1}, K^n) & \xrightarrow{d_r^{n+1}} & \pi_{r-1}(K^n) & \xrightarrow{j_{r-1}^n} & \pi_{r-1}(K^n, K^{n-1}) \\
 & & \downarrow & & & & \downarrow & &
\end{array}$$

We now define the homomorphisms, $\lambda_r^n : \Gamma_r^n \to \Gamma_r^{n+1}$, given by $\lambda_r^n(\alpha) = i_r^n(\alpha)$, $\alpha \in \Gamma_r^n$; $\mu_r^{n+1} : \Gamma_r^{n+1} \to H_r^n$; and $\nu_r^n : H_r^n \to \Gamma_{r-1}^{n-1}$; if $\beta \in \Gamma_r^{n+1}$ and $\beta = i_r^n(\gamma)$, $\gamma \in \pi_r(K^n)$, we define $\mu_r^{n+1}(\beta)$ as the homology class of $j_r^n(\gamma)$. It then follows, exactly as in the case $r=n$ already considered, that μ_r^{n+1} is single-valued. If $\alpha \in H_r^n$ and $z, \in \pi_r(K^n, K^{n-1})$, is a 'cycle' in the class α, we define $\nu_r^n(\alpha)$ as $d_r^n(z)$. We observe that, since $j_{r-1}^{n-1} d_r^n(z) = 0$, it follows that $d_r^n(z) \in (j_{r-1}^{n-1})^{-1}(0) = i_{r-1}^{n-2}\pi_{r-1}(K^{n-2})$, by exactness, $= \Gamma_{r-1}^{n-1}$; and also that if $z = j_r^n d_{r+1}^{n+1}(y)$, $y \in \pi_{r+1}(K^{n+1}, K^n)$, then

$$d_r^n(z) = d_r^n j_r^n d_{r+1}^{n+1}(y) = 0,$$

so that ν_r^n is, as asserted, a homomorphism of H_r^n into Γ_{r-1}^{n-1}.

We thus get the following 'derived' diagram:

$$\begin{array}{ccccc}
\Gamma_r^n & \xrightarrow{\mu_r^n} & H_r^{n-1} & \xrightarrow{\nu_r^{n-1}} & \Gamma_{r-1}^{n-2} \longrightarrow \\
\downarrow \lambda_r^n & & & & \downarrow \lambda_{r-1}^{n-2} \\
\Gamma_r^{n+1} & \xrightarrow{\mu_r^{n+1}} & H_r^n & \xrightarrow{\nu_r^n} & \Gamma_{r-1}^{n-1} \longrightarrow \\
\downarrow \lambda_r^{n+1} & & & & \downarrow \lambda_{r-1}^{n-1} \\
\Gamma_r^{n+2} & \xrightarrow{\mu_r^{n+2}} & H_r^{n+1} & \xrightarrow{\nu_r^{n+1}} & \Gamma_{r-1}^n \longrightarrow
\end{array}$$

THEOREM 2·7. *The sequence of groups and homomorphisms* $\ldots \to \Gamma_r^n \to H_r^{n-1} \to \Gamma_{r-1}^{n-2} \to \Gamma_{r-1}^{n-1} \to \ldots$ *is exact.*

We content ourselves with proving exactness at H_r^n, that is to say, we prove that $(\nu_r^n)^{-1}(0) = \mu_r^{n+1}\Gamma_r^{n+1}$.

Let $\alpha \in \Gamma_r^{n+1}$; then $\mu_r^{n+1}(\alpha)$ is the class of $j_r^n(\beta)$, $\beta \in \pi_r(K^n)$, where $\alpha = i_r^n(\beta)$. Thus, by definition, $\nu_r^n \mu_r^{n+1}(\alpha) = d_r^n j_r^n \beta = 0$. Conversely, let $\alpha \in H_r^n$ and let $\nu_r^n(\alpha) = 0$. Then $d_r^n(z) = 0$, where z is in the class α, so that $z = j_r^n(y)$, $y \in \pi_r(K^n)$, so that $\alpha = \mu_r^{n+1}(\beta)$, where $\beta = i_r^n(y)$.

Similar arguments establish the exactness of the sequence. Thus the situation in the derived diagram is algebraically analogous to that in the original diagram, and we may, by a similar construction, obtain the 'second derived' diagram, and so on. We thus get a sequence of such diagrams, called the *Massey homology spectrum*.† It is due to W. S. Massey. The derived diagrams are all homotopy invariants. Moreover, the subdiagram of the first derived diagram, consisting of groups Γ_r^n, $n \leqslant m$, and H_r^n, $n < m$, and related homomorphisms, is an invariant of m-type. A similar remark applies to the later derived diagrams.

If we extract from the first derived diagram the exact sequence 'passing through' H_n^n, and if we replace the symbols H_n^n, Γ_n^{n+1}, Γ_n^n by H_n, π_n, Γ_n, we get the exact sequence

$$\ldots \longrightarrow H_{n+1} \xrightarrow{\nu_{n+1}} \Gamma_n \xrightarrow{\lambda_n} \pi_n \xrightarrow{\mu_n} H_n \xrightarrow{\nu_n} \ldots, \qquad (2\cdot 8)$$

† The Massey spectrum provides an example of the application of the Leray-Koszul spectral homology theory (cf. J. P. Serre, *Ann. Math.* 54 (1951), 425–505, Chapter I).

where we also make suitable alterations in the names of the homomorphisms. This is the exact sequence which forms the subject-matter of the paper by J. H. C. Whitehead† called 'A certain exact sequence'.

The group π_n is the nth homotopy group of K, $\pi_n(K)$. If K is simply connected, H_n is the nth integral homology group of K, and μ_n is the natural homomorphism of $\pi_n(K)$ into $H_n(K)$. It will also be seen that a J_m-complex is precisely one in which $\Gamma_n = 0$, $n = 1, \ldots, m$, so that the exactness of the sequence provides an immediate proof of theorem 2·4. The homomorphism ν_n is called by Whitehead the *secondary boundary operator*.

If K is not simply connected, H_n may be interpreted as $H_n(\tilde{K})$, where $\tilde{K}$ is the universal covering complex of K. There are natural isomorphisms $\Gamma_n(\tilde{K}) \to \Gamma_n(K)$, $\pi_n(\tilde{K}) \to \pi_n(K)$, $n \geqslant 2$.

3. Realizability theorems. Suppose X and Y are any two arcwise-connected topological spaces and f is a mapping of X into Y. Then f induces homomorphisms $f_n^*: \pi_n(X) \to \pi_n(Y)$. We say that the homomorphisms f_n^* are *realized geometrically* by the map f. The importance of geometrical realizability has been underlined by results of J. H. C. Whitehead and others. In this section we will give a few of the basic results.‡

We first give an example of two spaces X and Y with isomorphic homotopy groups which are nevertheless not of the same homotopy type. For X we take the 2-sphere S^2; for Y we take the topological product of the 3-sphere S^3 and M, which is 'complex projective space of infinite dimensions', defined as follows.

We introduced in Chapter V, § 2, the space M_n, complex projective space of $(n-1)$ complex dimensions. We may embed M_k in M_n, $k < n$, by identifying the point $[z_1, z_2, \ldots, z_k]$ of M_k with the point $[z_1, z_2, \ldots, z_k, 0, \ldots, 0]$ of M_n. Then $M = \bigcup_n M_n$, and we

† See the reference at the end of the chapter. In this paper, Whitehead proves, for example, that if K is simply connected and dim $K \leqslant 4$, then the part of the sequence beginning with H_4 is an algebraic equivalent of homotopy type. He also proves a number of theorems on 'realizability' (see § 3 of this chapter).

‡ All these results are taken from CHI, but we do not state our theorems in the same generality.

give M the weak topology, so that $X \subset M$ is closed if and only if $X \cap M_n$ is closed for each n. It may be shown that M admits the cell-decomposition $e^0 \cup e^2 \cup e^4 \cup \ldots \cup e^{2n} \cup \ldots$, where

$$M^{2n} = M_{n+1},$$

and that e^{2n} is attached to M_n by the fibre-map $f: S^{2n-1} \rightarrow M_n$. Then $\pi_r(M) \approx \pi_r(M^{2r}) = \pi_r(M_{r+1})$, and, if $r \geqslant 3$, $\pi_r(M_{r+1}) \approx \pi_r(S^{2r+1})$, by theorem 2·6 of Chapter V. Thus $\pi_r(M) = 0$, $r \geqslant 3$. Also $\pi_1(M) = \pi_1(S^2) = 0$, and, again by theorem 2·6 of Chapter V, $\pi_2(M) \approx \pi_2(M_3) = Z_\infty$. Thus the homotopy groups of M are given by $\pi_r(M) = 0$, $r \neq 2$, $\pi_2(M) = Z_\infty$, a cyclic infinite group. If $Y = S^3 \times M$, then, by theorem 6·1 of Chapter IV,

$$\pi_r(Y) \approx \pi_r(S^3) + \pi_r(M).$$

Since, by theorem 2·1 of Chapter V, $\pi_r(S^3) \approx \pi_r(S^2)$, $r \geqslant 3$, and since $\pi_1(S^3) = \pi_1(S^2) = 0$ and $\pi_2(S^2) = Z_\infty$, it follows that, if $X = S^2$, then $\pi_r(X) \approx \pi_r(Y)$, all r. On the other hand, Y obviously possesses a non-bounding 3-cycle (the fundamental 3-cycle on S^3), so that the homology groups of X and Y are not isomorphic. Since the homology groups are invariants of homotopy type, it follows that X and Y furnish an example of the required kind.

In the example, the essential point is that the isomorphism of the homotopy groups of X and Y is 'abstract' and not susceptible of geometrical realization. This is brought out by the following theorem:

THEOREM 3·1. *If $f: K \rightarrow L$ is a map of the CW-complex K into the CW-complex L inducing isomorphisms $f_r^*: \pi_r(K) \approx \pi_r(L)$, $r = 1, 2, \ldots, \max(\dim K, \dim L)$, then f is a homotopy equivalence.*

(Note that this theorem includes the case where $\dim K$, or $\dim L$, $= \infty$.)

We give the proof of this theorem, because it introduces the concept of *mapping cylinder*, and demonstrates the great advantage deriving from the use of CW-complexes rather than simplicial complexes. However, we precede the proof by some remarks on the *identification topology*.

Let X be a topological space, Y an (untopologized) set of points, disjoint from X, and f a transformation of X on to Y. We say that a subset X_0 of X is *saturated* with respect to f if

$f^{-1}(fX_0)=X_0$. We then say that we are giving Y *the identification topology determined by* f when we specify the closed sets of Y as the images of saturated closed sets of X. This automatically makes f continuous. We note that, if f is to be continuous, then the closed sets of Y must be images of saturated closed sets of X. Thus we give Y the greatest number of closed sets consistent with the continuity of f. If X is a compact space, Y a Hausdorff space and f a map of X on to Y, then Y has the identification topology determined by f; for the image of any closed subset of X is closed in Y. Thus, if e^n is a cell of the cell-complex K, then $\bar{e}^n$ has the identification topology determined by the characteristic map $h: E^n \to \bar{e}^n$.

We say that a topology T on a set X is *weaker* than a topology T' (on the same set) if the closed sets of X under T' are also closed under T. Thus the weakest† topology on X is the discrete topology, and the strongest is that in which the only closed sets are X itself and the null set. With this definition, the identification topology may be characterized as the weakest topology consistent with the continuity of the map f. It should be noted that the definition gives significance to the definition of the *weak topology* of a cell-complex.

Now let f be a map‡ of X into Y. We form the space $(X \times I) \cup Y$, replacing Y by a suitable homeomorph if it is not already disjoint from $X \times I$. We also form the set $X \cup Y \cup (X \times (0,1))$, replacing the sets $X, Y, X \times (0,1)$ by homeomorphs if they are not already disjoint. We define a transformation g of $(X \times I) \cup Y$ on to $X \cup Y \cup X \times (0,1)$ by putting $g(x,0)=x, x \in X, g(x,1)=f(x)$, $g(x,t)=(x,t), t \in (0,1), g(y)=y, y \in Y$. If Z is the topological space obtained by giving the set $X \cup Y \cup (X \times (0,1))$ the identification topology determined by g, then we describe Z as *the mapping cylinder of the map* $f: X \to Y$. It is easily seen that X and Y retain their topologies as closed subsets of Z. From our point of view the utility of introducing the space Z springs from the fact that it contains X as a subspace and from the following theorem:

† It should be noted that this usage conflicts with that of analysis (e.g. weak convergence, strong convergence).

‡ Not necessarily *on to*.

THEOREM 3·2. *Y is a deformation retract of Z.*

(This certainly implies that the identity map $i: Y \rightarrow Z$ is a homotopy equivalence.)

The proof of this theorem depends on the following three lemmas:

LEMMA 3·3. *If Y has the identification topology determined by $f: X \rightarrow Y$ and if g is a (continuous) map of X into Z, then the transformation $gf^{-1}: Y \rightarrow Z$ is continuous if it is single-valued.*†

We prove first that if Z_0 is any subset of Z, then

$$(gf^{-1})^{-1}(Z_0) = f(g^{-1}(Z_0)).$$

For let $y \in (gf^{-1})^{-1}(Z_0)$. Then $gf^{-1}(y) = z \in Z_0$, so that $f^{-1}(y) \subset g^{-1}(z)$, or $y \in f(g^{-1}(z)) \subset f(g^{-1}(Z_0))$. Conversely, let $y \in f(g^{-1}(Z_0))$. Then $y = f(x), g(x) = z$, for some $x \in X, z \in Z_0$. Thus $x \in f^{-1}(y), z \in g(f^{-1}(y))$, so that $z = gf^{-1}(y)$, since gf^{-1} is single-valued. It follows that $y \in (gf^{-1})^{-1}(z) \subset (gf^{-1})^{-1}(Z_0)$.

This argument also shows that $g^{-1}(Z_0)$ is saturated with respect to f. For if $y \in (gf^{-1})^{-1}(Z_0)$, then $f^{-1}(y) \subset g^{-1}(Z_0)$. Thus

$$f^{-1}((gf^{-1})^{-1}(Z_0)), = f^{-1}(f(g^{-1}(Z_0))), \subset g^{-1}(Z_0),$$

whence $f^{-1}(f(g^{-1}(Z_0))) = g^{-1}(Z_0)$.

Now let Z_0 be closed in Z. Then $g^{-1}(Z_0)$ is a saturated closed set of X, so that $f(g^{-1}(Z_0))$ is closed in Y. Thus $(gf^{-1})^{-1}(Z_0)$ is closed if Z_0 is closed, whence gf^{-1} is continuous.

LEMMA 3·4. *If Y has the identification topology determined by $f: X \rightarrow Y$ and if $F: X \times I \rightarrow Y \times I$ is given by $F(x,t) = (f(x),t)$, $x \in X$, $t \in I$, then $Y \times I$ has the identification topology determined by F.*

We note first that the identification topology may also be characterized by the property that sets are *open* if and only if they are the images of saturated *open* sets (for if we maximize the closed sets we also maximize the open sets). It turns out to be convenient to use this characterization in the proof of the lemma.

Let G be a saturated open set of $X \times I$, so that $F^{-1}(F(G)) = G$. We wish to show that $F(G)$ is open in $Y \times I$. Let $(y_0, t_0) \in F(G)$ and let $(x_0, t_0) \in G$ be such that $f(x_0) = y_0$. Let T stand for the (non-empty) set of all $t \in I$ such that $(x_0, t) \in G$. T is obviously an

† We used a special case of this lemma in the proof of lemma 1·5. The X, Y, Z of the lemmas are not, of course, the same as those of 3·2.

open set containing t_0, and we may choose an open set T_0 such that $t_0 \in T_0, \bar{T}_0 \subset T$. Then $x_0 \times \bar{T}_0 \subset G$. We may find an open set $V(x_0)$, containing x_0, and such that $V(x_0) \times \bar{T}_0 \subset G$. For we may choose, for each $t' \in \bar{T}_0$, open sets† $V'(x_0)$, $U(t')$ such that $x_0 \in V'(x_0)$, $t' \in U(t')$ and $V'(x_0) \times U(t') \subset G$. Since $\bar{T}_0$ is compact, a finite number of the sets $\{U(t')\}$, say $U(t_1), \ldots, U(t_n)$, covers $\bar{T}_0$. If the corresponding neighbourhoods of x_0 are $V_1(x_0), \ldots, V_n(x_0)$, then $V_1(x_0) \cap \ldots \cap V_n(x_0)$ is an open set $V(x_0)$, containing x_0 and such that $V(x_0) \times \bar{T}_0 \subset G$.

Now
$$F(V(x_0) \times \bar{T}_0) = fV(x_0) \times \bar{T}_0,$$
and
$$F^{-1}(F(V(x_0) \times \bar{T}_0)) = f^{-1}(fV(x_0)) \times \bar{T}_0.$$
Since $V(x_0) \times \bar{T}_0 \subset G$, $F^{-1}(F(V(x_0) \times \bar{T}_0)) \subset F^{-1}(F(G)) = G$, since G is saturated. Thus $f^{-1}(fV(x_0)) \times \bar{T}_0 \subset G$.

Let us define $W(x_0)$ as the largest open set, containing x_0, such that $W(x_0) \times \bar{T}_0 \subset G$. More precisely, $W(x_0)$ is the union of all neighbourhoods, $V(x_0)$, of x_0, with the property that
$$V(x_0) \times \bar{T}_0 \subset G.$$
We prove that $f^{-1}(fW(x_0)) = W(x_0)$. For, if not, let
$$x \in f^{-1}(fW(x_0)), \quad x \notin W(x_0).$$
Then $x \times \bar{T}_0 \subset G$, so that there exists an open set $V(x)$, containing x, and such that $V(x) \times \bar{T}_0 \subset G$. Then $(W(x_0) \cup V(x)) \times \bar{T}_0 \subset G$, contradicting the definition of $W(x_0)$. Thus $W(x_0)$ is a saturated open set in X, so that $fW(x_0)$ is open in Y. Since $W(x_0) \times T_0 \subset G$, $fW(x_0) \times T_0$, $= F(W(x_0) \times T_0)$, $\subset F(G)$, and is an open set containing (y_0, t_0). Thus $F(G)$ is open and the lemma is proved.

LEMMA 3·5. *If Y has the identification topology determined by $f: X \to Y$ and if g_t is a homotopy of maps of X into Z, then the homotopy $g_t f^{-1}: Y \to Z$ is continuous if it is single-valued.*

This is an immediate consequence of lemmas 3·3 and 3·4.

We now return to the proof of theorem 3·2. Let
$$\theta_t: (X \times I) \cup Y \to (X \times I) \cup Y$$
be the deformation given by $\theta_t \mid Y = 1$, $\theta_t(x, t') = (x, t + t' - tt')$,

† The open sets $U(t')$ need not be contained in $\bar{T}_0$, though there is no harm in insisting that they are.

$x \in X$, $t' \in I$. Given the identification map $g: X \times I \cup Y \to Z$, define the homotopy $\eta_t: Z \to Z$ by $\eta_t = g\theta_t g^{-1}$. Then η_t is single-valued. For θ_t is the identity on $(X \times 1) \cup Y$, so that η_t is the identity on Y (as a subset of Z); $\eta_t(x) = (x, t)$, $0 < t < 1$, $\eta_0(x) = x$, $\eta_1(x) = f(x)$, $x \in X$ (as a subset of Z); and $\eta_t(x, t') = (x, t + t' - tt')$, $0 \leqslant t < 1$, $0 < t' < 1$, $\eta_1(x, t') = f(x)$, $0 < t' < 1$. Thus, by lemma 3·5, η_t is continuous. Clearly $\eta_0 = 1$, $\eta_t \mid Y = 1$, $\eta_1 Z = Y$. Thus Y is a deformation retract of Z, and η_t is a retracting deformation.

We use the device of the mapping cylinder to prove theorem 3·1. Here we specialize the spaces X, Y to CW-complexes K, L and we retain the symbol Z for the mapping cylinder of $f: K \to L$. We have shown that L is a deformation retract of Z and that η_t is a deformation. We wish to show that, under the special assumptions of the theorem, the identity map $i: K \to Z$ is a homotopy equivalence.

We may, without loss of generality, assume that the map f is cellular (for, otherwise, we would merely replace it by a homotopic map which was cellular). Then Z can be given the structure of a cell-complex, whose cells are those of K and those of L and the cells $e^n \times (0, 1)$, where e^n is a cell of K. $K \times I$ and L being CW-complexes, it is not difficult† to show that Z is a CW-complex, and it is this crucial fact which serves, in this context, to justify the introduction of CW-complexes.

Now let $i: K \to Z$ be the identity map. Then, defining $\rho: Z \to L$ by $\rho(z) = \eta_1(z)$, $z \in Z$, we see that $\rho i: K \to L$ is simply the map f. Since ρ is a homotopy equivalence, and since f induces isomorphisms $f_r^*: \pi_r(K) \approx \pi_r(L)$, $r = 1, 2, \ldots, N = \max(\dim K, \dim L)$, it follows‡ that i induces isomorphisms $i_r: \pi_r(K) \approx \pi_r(Z)$, $r = 1, 2, \ldots, N$. It follows from the exactness of the homotopy sequence of the pair (Z, K) that $\pi_r(Z, K) = 0$, $r = 1, \ldots, N$. Hence the identity map $Z \to Z$ is homotopic,§ rel K, to a map $h: Z \to Z$ such that $h(Z^N) \subset K$. If $\dim K < \dim L$ (or if $\dim K = \infty$), h is itself a retraction of Z on to K. Otherwise $\dim K = N$ and $\dim Z = N + 1$, the

† If K, L were simplicial complexes, the triangulation of Z would present tiresome difficulties. For the proof that Z is a CW-complex, see CHI, p. 235.

‡ Recall that we allow $N = \infty$

§ By theorem 1·7 and the remark subsequent to it. Note that

$$\dim Z = \max((\dim K) + 1, \dim L).$$

$(N+1)$-cells of Z being the cells $e^N \times (0,1)$, $e^N \subset K$. Consider the cell $e^N \times (0,1)$. Let $k: E^{N+1}, \dot{E}^{N+1} \to Z, Z^N$ be a characteristic map for this cell, and consider the map $hk: E^{N+1}, \dot{E}^{N+1} \to Z, K$. Since i_N maps $\pi_N(K)$ *isomorphically* on to $\pi_N(Z)$, and since the map $hk \mid \dot{E}^{N+1}$ admits an extension to a map of E^{N+1} into Z, it follows that $hk \mid \dot{E}^{N+1}$ admits an extension to a map of E^{N+1} into K, say, to $s: E^{N+1} \to K$. Define $h^*: Z^N \cup e^N \times (0,1) \to K$ by $h^* \mid Z^N = h$, $h^* \mid \overline{e^N \times (0,1)} = sk^{-1}$. It is clear that $h^* \mid \overline{e^N \times (0,1)}$ is single-valued and hence (by lemma 3·3, for instance) continuous. We may extend $h \mid Z^N$ in this way over all the $(N+1)$-cells of Z, obtaining a retraction $h^*: Z \to K$, which is continuous by theorem 1·3. Note that $h^* \mid K = 1$.

Again, since $\pi_r(Z, K) = 0$, $r = 1, \ldots, N$, the identity map $j: L \to Z$ is deformable into $j': L \to K$. Thus if we first deform Z on to L by η_t, and then deform L into K, we get a deformation of Z into K. Thus

$$\eta_0 \sim \eta_1 = j\rho \sim j'\rho = ih^*j'\rho \sim ih^*: Z \to Z.$$

Since $h^* \mid K = 1$, and, of course, $\eta_0 = 1$, this shows that $i: K \to Z$ is a homotopy equivalence. Thus $\rho i = f: K \to L$ is a homotopy equivalence, and theorem 3·1 is proved. It may be noted that $h^*j: L \to K$ is a homotopy inverse of f. It is also worth observing that the proof of the theorem would have been rendered somewhat easier if we had been given that the map $f: K \to L$ induces isomorphisms of $\pi_r(K)$ on to $\pi_r(L)$, $r = 1, \ldots, M-1$, and a homomorphism of $\pi_M(K)$ on to $\pi_M(L)$, $M = \max((\dim K) + 1, \dim L)$.

THEOREM 3·6. *If $f: K \to L$ induces $f_r: \pi_r(K) \approx \pi_r(L)$, $r = 1, \ldots, N-1$, $N = \max(\dim K, \dim L)$, then f is an N-equivalence.*

This is proved very similarly to theorem 3·1, from which it differs only if $N < \infty$. Assuming this, and using the notation of the proof of that theorem, let $Z_0^N \subset Z$ be the image of

$$(K^{N-1} \times I) \cup L$$

under the identification map $g: K \times I \cup L \to Z$. Since

$$(K^{N-1} \times I) \cup (K \times I)$$

is a retract of $K \times I$ it follows that Z_0^N is a retract† of Z. Let θ be the retraction. Now we may prove, as before, that there exists $\rho: Z^N \to K$ with $\rho \mid K = 1$ and‡ $\rho \mid Z^{N-1} \sim 1 : Z^{N-1} \to Z^N$. Then $\rho\theta: Z \to K$ is an N-inverse of $i: K \to Z$. For

$$\rho\theta i \mid K^{N-1} = 1, \quad i\rho\theta \mid Z^{N-1} = i\rho \mid Z^{N-1} \sim 1 \mid Z^{N-1}.$$

Thus $i: K \to Z$ is an N-equivalence, whence $f: K \to L$ is an N-equivalence.

Now let $f: K \to L$ be a cellular map and choose vertices $x_0 \in K$, $y_0 \in L$ with $f(x_0) = y_0$. Let $\tilde{K}, \tilde{L}$ be the universal covering complexes of K, L and choose $\tilde{x}_0, \tilde{y}_0$ lying over x_0, y_0 respectively. Then there is a unique map $\tilde{f}: \tilde{K} \to \tilde{L}$ such that $p\tilde{f} = fp$ and $\tilde{f}\tilde{x}_0 = \tilde{y}_0$, where we use p to denote the projections $\tilde{K} \to K$, $\tilde{L} \to L$. Let $\tilde{f}$ induce $\tilde{f}_r: H_r(\tilde{K}) \to H_r(\tilde{L})$. Then we say that $\tilde{f}_r$ is induced by $f: K \to L$.

THEOREM 3·7. *If $f: K \to L$ induces*

$$f_1: \pi_1(K) \approx \pi_1(L), \quad \tilde{f}_r: H_r(\tilde{K}) \approx H_r(\tilde{L}), \quad r = 2, 3, \ldots,$$

then f is a homotopy equivalence.

The injections $K \to Z$, $L \to Z$ induce isomorphisms

$$\pi_1(K) \approx \pi_1(Z), \quad \pi_1(L) \approx \pi_1(Z).$$

Thus if $\tilde{Z}$ is the universal cover of Z, it may be regarded as containing $\tilde{K}$ and $\tilde{L}$. Then using the homology sequence of the pair $\tilde{Z}, \tilde{K}$ we may show that $H_n(\tilde{Z}, \tilde{K}) = 0$ for all n, whence, by theorem 3·3 of Chapter III, $\pi_n(\tilde{Z}, \tilde{K}) = 0$, all n. By theorem 1·8 of Chapter V, this implies that $\pi_n(Z, K) = 0$, all n; from this point the proof proceeds as for theorem 3·1.

THEOREM 3·8. *If K, L are simply-connected, and if $f: K \to L$ induces $f_r: H_r(K) \approx H_r(L)$, $r = 2, 3, \ldots$, then f is a homotopy equivalence.*

For in this case K and L are their own universal covers.

For further results on realizability, see J. H. C. Whitehead, 'Combinatorial homotopy, II', *Bull. Amer. Math. Soc.* 55, 5 (1949), 453–96; 'A certain exact sequence', *Ann. Math.* 52, 1 (1950), 51–110.

† If $\sigma: (K \times I) \cup L \to (K^{N-1} \times I) \cup (K \times I) \cup L$ is a retraction, then

$$g\sigma g^{-1}: Z \to Z_0^N$$

is a retraction. The continuity of $\theta = g\sigma g^{-1}$ follows, of course, from lemma 3·3.

‡ It would be more correct to say that $i\rho \mid Z^{N-1} \sim 1 : Z^{N-1} \to Z^N$ as ρ is a map into K. This point is brought out in the next line.

CHAPTER VIII

HOMOTOPY GROUPS OF COMPLEXES

1. Statement of the problem. We study in this final chapter a particularly simple case of a particular problem in homotopy theory. The problem is to calculate the homotopy groups of CW-complexes; and since such calculations must be based on a knowledge of the homotopy groups of spheres, we restrict ourselves to cases in which the relevant homotopy groups of spheres are known. In fact, the main object of our study will be the determination of $\pi_{n+1}(K)$ in terms of the homology characteristics of K, where K is a finite cell-complex whose first $(n-1)$ homotopy groups vanish. We will, moreover, take $n>2$ for simplicity. The more difficult case $n=2$ was solved by J. H. C. Whitehead in his paper 'A certain exact sequence', *Ann. Math.* 52, 1 (1950), 51–110. We base our approach on the methods of that paper, merely making the simplifications arising from the fact that we take $n>2$; but we also give a slightly different approach, due to the author,† but essentially inspired by an earlier paper of J. H. C. Whitehead, 'The homotopy type of a special kind of polyhedron', *Ann. Soc. Polon. Math.* 21 (1948), 176–86. Finally, we indicate possible extensions of the method to the calculation of the higher homotopy groups of K.

We wish to stress that this 'example' of the application of homotopy theory is chosen because of the author's familiarity with it, and not because it necessarily occupies a central position in homotopy theory. However, the introduction into topology of homotopy groups takes on a further justification if rules can be given for calculating them, so that a motivation is provided for the work described in this chapter.

† See *Quart. J. Math. Oxford* (2), 1 (1950), 299–309. Following J. H. C. Whitehead, the author referred to a complex having the properties of K as an A_n^2-polyhedron.

2. Whitehead's exact sequence. We recall from Chapter VII (2·8) the exact sequence of a simply-connected CW-complex, namely,

$$\ldots \longrightarrow H_{r+1}(K) \xrightarrow{\nu_{r+1}} \Gamma_r(K) \xrightarrow{\lambda_r} \pi_r(K) \xrightarrow{\mu_r} H_r(K) \longrightarrow \ldots. \quad (2{\cdot}1)$$

We also recall that $\Gamma_r(K) = i_r \pi_r(K^{r-1}) \subset \pi_r(K^r)$. We take $\pi_r(K^r, K^{r-1})$ as the rth chain group of K. The boundary operator $\delta_r : \pi_r(K^r, K^{r-1}) \to \pi_{r-1}(K^{r-1}, K^{r-2})$ is $j_{r-1} d_r$, $r > 2$ (and d_r, $r = 2$), where d_r is the homotopy boundary $d_r : \pi_r(K^r, K^{r-1}) \to \pi_{r-1}(K^{r-1})$ and j_{r-1} is the injection $j_{r-1} : \pi_{r-1}(K^{r-1}) \to \pi_{r-1}(K^{r-1}, K^{r-2})$. Finally, we recall from Chapter VII the definitions of ν_{r+1}, λ_r, and μ_r.

Let $z \in Z_{r+1}(K)$, the group of $(r+1)$-cycles, and let $[z]$ be its homology class. Then $d_{r+1}(z) \in i_r \pi_r(K^{r-1}), = \Gamma_r(K)$, and we define $\nu_{r+1}[z] = d_{r+1}(z)$. If $x \in \Gamma_r(K)$, then $\lambda_r(x)$ is the image of x under the injection $k_r : \pi_r(K^r) \to \pi_r(K)$. Given $y \in \pi_r(K)$, let $y' \in \pi_r(K^r)$ be chosen in $k_r^{-1}(y)$. Then $\mu_r(y) = [j_r(y')]$.

We assume $\pi_r(K) = 0$, $r = 1, \ldots, n-1$. Then, as shown in Chapter VII, $\mu_n : \pi_n(K) \approx H_n(K)$ and μ_{n+1} maps $\pi_{n+1}(K)$ on to $H_{n+1}(K)$. We extract from (2·1) the subsequence

$$H_{n+2}(K) \xrightarrow{\nu_{n+2}} \Gamma_{n+1}(K) \xrightarrow{\lambda_{n+1}} \pi_{n+1}(K) \xrightarrow{\mu_{n+1}} H_{n+1}(K) \to 0. \quad (2{\cdot}2)$$

THEOREM 2·3. *$\pi_{n+1}(K)$ is an extension*† *of $H_{n+1}(K)$ by*

$$\Gamma_{n+1}(K) - \nu_{n+2} H_{n+2}(K).$$

For μ_{n+1} maps $\pi_{n+1}(K)$ on to $H_{n+1}(K)$ with kernel $\lambda_{n+1} \Gamma_{n+1}(K)$. Now λ_{n+1} maps $\Gamma_{n+1}(K)$ on to $\lambda_{n+1} \Gamma_{n+1}(K)$ with kernel $\nu_{n+2} H_{n+2}(K)$, so that λ_{n+1} induces an isomorphism

$$(\Gamma_{n+1}(K) - \nu_{n+2} H_{n+2}(K)) \approx \lambda_{n+1} \Gamma_{n+1}(K).$$

To calculate $\pi_{n+1}(K)$ in terms of the homology characters of K, it remains to elucidate the group $\Gamma_{n+1}(K)$ and the homomorphism ν_{n+2} and to describe the group extension of $H_{n+1}(K)$. At this stage we introduce the assumption $n > 2$.

Let η be the generating element of $\pi_{n+1}(S^n)$. Then, since η is a suspension element, it follows from theorem 2·3 of Chapter VI

† The group K is an extension of the group H by the group G if there is a homomorphism of K on to H with kernel isomorphic to G.

that the mapping $\alpha \to \alpha \circ \eta$, $\alpha \in \pi_n(K)$, induces a homomorphism $\bar{\eta}: \pi_n(K) \to \pi_{n+1}(K)$. However, since any $\alpha \in \pi_n(K)$ has a representative map $f: S^n \to K^n$, it follows that $\alpha \to \alpha \circ \eta$ in fact determines a homomorphism $\theta: \pi_n(K) \to i_{n+1}\pi_{n+1}(K^n) = \Gamma_{n+1}(K)$.

THEOREM 2·4. *θ is a homomorphism of $\pi_n(K)$ on to $\Gamma_{n+1}(K)$ with kernel $2\pi_n(K)$.*

Note first that, since $2\eta = 0$, $2\alpha \circ \eta = \alpha \circ 2\eta = 0$, so that $2\pi_n(K)$ certainly lies in the kernel.

Now $\pi_n(K^n) \approx H_n(K^n)$, the group of n-cycles, so that $\pi_n(K^n)$ is free Abelian. Let $\{a_i\}$ be a set of free generators of $\pi_n(K^n)$, let $\{e_\lambda^{n+1}\}$ be the $(n+1)$-cells of K and let $c_\lambda \in \pi_{n+1}(K^{n+1}, K^n)$ be the element represented by a characteristic map for e_λ^{n+1}. Then $\{c_\lambda\}$ is a set of free generators of $\pi_{n+1}(K^{n+1}, K^n)$ and, since $\pi_n(K) \approx \pi_n(K^n) - d_{n+1}\pi_{n+1}(K^{n+1}, K^n)$, $\pi_n(K)$ is generated by $\{a_i\}$ with relations $b_\lambda = d_{n+1}c_\lambda = 0$.

Let $K_0^n = e^0 \cup \{e_i^n\}$, where $\{e_i^n\}$ is a set of n-cells† in (1-1) correspondence with the $\{a_i\}$. Then $e^0 \cup e_i^n$ is an n-sphere S_i^n and $\pi_n(K_0^n)$ is freely generated by the set of elements $\{a_i^0\}$, where a_i^0 is represented by a homeomorphism $\phi_i: S^n \to S_i^n$. Let $\psi: K_0^n \to K^n$ be such that $\psi\phi_i$ represents a_i. Since $\pi_r(K_0^n) = \pi_r(K^n) = 0$, $r < n$, and since ψ obviously induces the isomorphism $\pi_n(K_0^n) \approx \pi_n(K^n)$ given by $a_i^0 \leftrightarrow a_i$, it follows from theorem 3·1 of Chapter VII that ψ is a homotopy equivalence. Thus, in particular, ψ induces $\psi_{n+1}: \pi_{n+1}(K_0^n) \approx \pi_{n+1}(K^n)$. Now if $n > 2$, $n+1 < 2n-1$, so that it follows from (4·3) of Chapter VI that

$$\pi_{n+1}(S_1^n \cup S_2^n) \approx \pi_{n+1}(S_1^n) + \pi_{n+1}(S_2^n), \quad n > 2, \tag{2·5}$$

where $S_1^n \cup S_2^n$ is the union of two n-spheres with a single common point. By an obvious extension of the argument used in Chapter VI, we may extend (2·5) to a similar formula for the union of any (finite) number of n-spheres; thus

$$\pi_{n+1}(S_1^n \cup \ldots \cup S_k^n) \approx \pi_{n+1}(S_1^n) + \ldots + \pi_{n+1}(S_k^n), \quad n > 2. \tag{2·6}$$

Now since any compact subset of K_0^n is contained in the union of a finite number of the S_i^n, and since this union is a retract of K_0^n, it follows from (2·6) that any element in $\pi_{n+1}(K_0^n)$ is uniquely

† Possibly infinite in number, but mutually disjoint.

expressible as a (finite) sum of elements $a_i^0 \circ \eta$. In other words, $\pi_{n+1}(K_0^n)$ is a free mod 2 module, freely generated by $\{a_i^0 \circ \eta\}$. Clearly $\psi_{n+1}(a_i^0 \circ \eta) = a_i \circ \eta$, so that $\pi_{n+1}(K^n)$ is a free mod 2 module freely generated by $\{a_i \circ \eta\}$. Since

$$\Gamma_{n+1}(K) = i_{n+1}\pi_{n+1}(K^n),$$
$$\subset \pi_{n+1}(K^{n+1}), \approx \pi_{n+1}(K^n) - d\pi_{n+2}(K^{n+1}, K^n),$$

the theorem will be proved when we have shown that the relations $d\pi_{n+2}(K^{n+1}, K^n) = 0$ are precisely the relations $b_\lambda \circ \eta = 0$.

Let P be a CW-complex which is the union of $(n+1)$-elements E_λ^{n+1} with a single common point (on the boundary of each E_λ^{n+1}), where the E_λ^{n+1} are in (1-1) correspondence with the $(n+1)$-cells, e_λ^{n+1}, of K. Let Q be the union of the boundary n-spheres, $\dot{E}_\lambda^{n+1}$, and let $h: P, Q \to K^{n+1}, K^n$ be such that $h \mid E_\lambda^{n+1}$ is a characteristic map for e_λ^{n+1}. Let h induce $g_{n+2}: \pi_{n+2}(P, Q) \to \pi_{n+2}(K^{n+1}, K^n)$. By Whitehead's suspension theorem,†

$$\pi_{n+2}(K^{n+1}, K^n) = g_{n+2}\pi_{n+2}(P, Q).$$

Let $h \mid Q$ induce $h_{n+1}: \pi_{n+1}(Q) \to \pi_{n+1}(K^n)$ and consider the diagram

$$\begin{array}{ccc} \pi_{n+2}(P, Q) & \xrightarrow{d} & \pi_{n+1}(Q) \\ \downarrow{\scriptstyle g_{n+2}} & & \downarrow{\scriptstyle h_{n+1}} \\ \pi_{n+2}(K^{n+1}, K^n) & \xrightarrow{d} & \pi_{n+1}(K^n) \end{array}$$

where the same symbol d stands for both boundary homomorphisms. Since P is contractible, we have

$$d: \pi_{n+2}(P, Q) \approx \pi_{n+1}(Q),$$

and since the 'commutative law', $dg_{n+2} = h_{n+1}d$, obviously holds, we have

$$d\pi_{n+2}(K^{n+1}, K^n) = dg_{n+2}\pi_{n+2}(P, Q)$$
$$= h_{n+1}d\pi_{n+2}(P, Q) = h_{n+1}\pi_{n+1}(Q).$$

Now, by definition, $(h \mid \dot{E}_\lambda^{n+1})\,\phi_\lambda$ represents $b_\lambda \in \pi_n(K^n)$, where ϕ_λ is a homeomorphism $S^n \to \dot{E}_\lambda^{n+1}$. Also $\pi_n(Q)$ is freely generated

† Theorem 1 of 'A note on suspension', *Quart. J. Math. Oxford* (2), 1 (1950), 9–22. In fact, g_{n+2} is isomorphic. See also 2.11 and 2.12 of Chapter VI.

by $\{e_\lambda\}$, where e_λ is represented by ϕ_λ, and, as proved earlier for K_0^n, $\pi_{n+1}(Q)$ is a free mod 2 module freely generated by $\{e_\lambda \circ \eta\}$. Clearly $h_{n+1}(e_\lambda \circ \eta) = b_\lambda \circ \eta$, so that the relations

$$d\pi_{n+2}(K^{n+1}, K^n) = 0$$

are, as asserted, just the relations $b_\lambda \circ \eta = 0$, and the theorem is proved.

Since $\pi_n(K) \approx H_n(K)$, we have $\Gamma_{n+1}(K) \approx (H_n(K))_2$, i.e. $H_n(K)$ reduced mod 2, and ν_{n+2} may then be identified with a homomorphism† $\nu_{n+2}: H_{n+2}(K) \to (H_n(K))_2$. Thus if we are given $H_n(K)$, $H_{n+1}(K)$, $H_{n+2}(K)$, and ν_{n+2}, we know that $\pi_{n+1}(K)$ is an extension of a given group by a given group. It remains to calculate this extension. Since we are interested in $\pi_{n+1}(K)$, it will be sufficient to replace K by K^{n+2} if K is, in fact, of dimension $> n+2$.

We now define‡ $H_{n+2}(2)$ as $(\delta_{n+2}^{-1}(2Z_{n+1}(K)))_2$, where $Z_{n+1}(K)$ is the group of $(n+1)$-cycles of K, and will define a homomorphism $\nu(2): H_{n+2}(2) \to \Gamma_{n+1}(K)$. Since $\Gamma_{n+1}(K)$ is a mod 2 module, and $H_{n+2}(2)$ is just $\delta_{n+2}^{-1}(2Z_{n+1}(K))$ reduced mod 2, it is obviously sufficient to define a homomorphism

$$\bar{\nu}: \delta_{n+2}^{-1}(2Z_{n+1}(K)) \to \Gamma_{n+1}(K).$$

Let $c \in \delta_{n+2}^{-1}(2Z_{n+1}(K))$. Then $\delta_{n+2}(c) = 2z$, $z \in Z_{n+1}(K)$. We now point out that $Z_{n+1}(K) = j_{n+1}\pi_{n+1}(K^{n+1})$. For, since $\Gamma_n(K) = 0$, it follows that j_n maps $\pi_n(K^n)$ isomorphically into $\pi_n(K^n, K^{n-1})$, so that

$$\begin{aligned} Z_{n+1}(K) = \delta_{n+1}^{-1}(0) &= (j_n d_{n+1})^{-1}(0) \\ &= d_{n+1}^{-1}(j_n^{-1}(0)) = d_{n+1}^{-1}(0) = j_{n+1}\pi_{n+1}(K^{n+1}). \end{aligned}$$

Thus $z = j_{n+1}(y)$, $y \in \pi_{n+1}(K^{n+1})$, and $\delta_{n+2}(c) = j_{n+1}(2y)$. Thus $j_{n+1}(d_{n+2}(c) - 2y) = 0$, so that $d_{n+2}c - 2y \in \Gamma_{n+1}(K)$. Moreover, if $z' = j_{n+1}y'$ and $2z' = 2z = \delta_{n+2}c$, then it follows from the fact $C_{n+1}(K)$ is free Abelian that $z' = z$, so that $j_{n+1}(y' - y) = 0$, whence

† Recall that the homomorphism ν is called the 'secondary boundary' by J. H. C. Whitehead.

‡ The group $H_{n+2}(2)$ is just the $(n+2)$nd homology group of K with coefficients the integers reduced mod 2. If dim $K > n+2$ we would have

$$H_{n+2}(2) = (\delta_{n+2}^{-1}(2Z_{n+1}(K))/\delta_{n+3}C_{n+3}(K))_2.$$

See *A certain exact sequence*, p. 55.

$y'-y\in\Gamma_{n+1}(K)$ and $2y'=2y$, since every non-zero element in $\Gamma_{n+1}(K)$ is of order 2. Thus the mapping $c\to d_{n+2}c-2y$ is single-valued and it is obviously a homomorphism. If we call this homomorphism $\bar{\nu}$, then it is clear that $\bar{\nu}\,|\,Z_{n+2}(K)$ is precisely ν_{n+2}, when we make the natural identification

$$Z_{n+2}(K)\approx H_{n+2}(K),$$

K being $(n+2)$-dimensional. Since $\nu(2)$ determines and is determined by $\bar{\nu}$, it follows that knowledge of $\nu(2)$ implies knowledge of ν_{n+2}.

We will show that $\nu(2)$ determines $\pi_{n+1}(K)$ as a group extension if K is finite. Then $H_{n+1}(K)=B+T_o+T_e$, where B is free Abelian, T_o is the direct sum of cyclic groups of odd order and T_e is the direct sum of cyclic groups of even order. Let $(b_1,\ldots,b_\rho)$ be a free basis for B, let T_o be generated by $(v_1,\ldots,v_\sigma)$ of odd orders $\tau_1,\ldots,\tau_\sigma$, and let $a_1,\ldots,a_\rho,a'_1,\ldots,a'_\sigma$ be elements of $\pi_{n+1}(K)$ such that $\mu_{n+1}(a_1)=b_1,\ldots,\mu_{n+1}(a_\rho)=b_\rho$, $\mu_{n+1}(a'_1)=v_1,\ldots,\mu_{n+1}(a'_\sigma)=v_\sigma$. Regard T_o as generated by $(2v_1,\ldots,2v_\sigma)$ and define the homomorphism $\mu^*:B+T_o\to\pi_{n+1}(K)$ by

$$\mu^*(b_i)=a_i,\quad i=1,\ldots,\rho;$$

$$\mu^*(2v_j)=2a'_j,\quad j=1,\ldots,\sigma.$$

Then μ^* is single-valued† since $\mu^*(2\tau_j v_j)=2\tau_j a'_j$; but

$$\tau_j a'_j\in\mu^{-1}_{n+1}(0),$$

and $\mu^{-1}_{n+1}(0)$ is a mod 2 module, so that $\mu^*(2\tau_j v_j)=0$. Also $\mu_{n+1}\mu^*$ is the identity. Thus $\pi_{n+1}(K)=\mu^*(B+T_o)+\pi^e$, where μ^* is an isomorphism and π^e is an extension of T_e by

$$\Gamma_{n+1}(K)-\nu_{n+2}H_{n+2}(K).$$

It remains to determine π^e. Let T_e be generated by $(t_1,\ldots,t_m)$ of (even) orders $\tau_1,\ldots,\tau_m$. Let $(c_1,\ldots,c_m,c_{m+1},\ldots)$ be a canonical basis for $C_{n+2}(K)$, and let $\delta c_i=\tau_i z_i$, $i=1,\ldots,m$, where z_i is an $(n+1)$-cycle in the class t_i. Now $z_i=j_{n+1}(y_i)$, $y_i\in\pi_{n+1}(K^{n+1})$, and $k_{n+1}(y_i),\in\pi_{n+1}(K)$, is a representative in π^e of t_i, where k_{n+1}

† We regard v_j as $\dfrac{1+\tau_j}{2}(2v_j)$; thus we might have defined μ^* on T_o by $\mu^*(v_j)=(1+\tau_j)\,a'_j$. This is strictly speaking a different μ^* from the one given in the text, but would serve as well.

is, as before, the injection $k_{n+1}: \pi_{n+1}(K^{n+1}) \to \pi_{n+1}(K)$. We will compute $\tau_i(k_{n+1}(y_i)) \in \Gamma_{n+1}(K) - \nu_{n+2} H_{n+2}(K)$. Now

$$\bar{\nu}(c_i) = d_{n+2}(c_i) - \tau_i y_i,$$

so that $k_{n+1}(\tau_i y_i) = -k_{n+1} \bar{\nu}(c_i)$. Now

$$\bar{\nu}(c_i) \in \Gamma_{n+1}(K) \quad \text{and} \quad k_{n+1} \mid \Gamma_{n+1}(K)$$

is precisely the homomorphism λ_{n+1} of (2·2); and we have identified $\lambda_{n+1} \Gamma_{n+1}(K)$ with its isomorph $\Gamma_{n+1}(K) - \nu_{n+2} H_{n+2}(K)$. Thus† $\tau_i(k_{n+1}(y_i)) = k_{n+1}(\tau_i y_i) = \bar{k}\bar{\nu}(c_i)$, where $\bar{k}$ is the natural homomorphism $\Gamma_{n+1}(K) \to \Gamma_{n+1}(K) - \nu_{n+2} H_{n+2}(K)$. The elements $\tau_i(k_{n+1} y_i)$ determine, up to equivalence, the extension π^e. Since $\bar{k}$ is determined by ν_{n+2}, and both ν_{n+2} and $\bar{\nu}$ are determined by $\nu(2)$, we have shown how $\nu(2)$ determines $\pi_{n+1}(K)$ as a group extension.

3. The homology system and the reduced complex. Let us identify $\Gamma_{n+1}(K)$ with $H_n(K)/2H_n(K)$ by means of the homomorphism θ of theorem 2·4 and the natural isomorphism $\pi_n(K) \approx H_n(K)$, where K is a finite cell-complex, of dimension $\leqslant n+2$, whose first $(n-1)$ homotopy groups vanish. Then we have seen that the sequence (2·2) is determined, up to isomorphism, by the groups $H_n(K)$, $H_{n+1}(K)$, $H_{n+2}(K)$, and $H_{n+2}(2)$ and the homomorphism $\nu(2): H_{n+2}(2) \to H_n(K)/2H_n(K)$, provided, of course, that $n > 2$.

Following Whitehead, the author has defined an (abstract) *homology system*‡ as follows. It consists of the finitely generated abstract groups H_n, H_{n+1}, H_{n+2}, $H_{n+2}(2)$, together with certain homomorphisms. The groups are all Abelian and H_{n+2} is free Abelian. A homomorphism $\Delta: H_{n+2}(2) \to H_{n+1}$ is given which is§ on to ${}_2H_{n+1}$ and which admits a right inverse $\Delta^*: {}_2H_{n+1} \to H_{n+2}(2)$. A homomorphism $\mu: H_{n+2} \to H_{n+2}(2)$ maps each element of H_{n+2}

† The minus sign in $-k_{n+1}\bar{\nu}(c_i)$ may, of course, be suppressed.

‡ Actually, Whitehead defined a cohomology system. This is formally identical with a homology system, but was related to the cohomology groups of a complex, instead of the homology groups.

§ Given an Abelian group G and an integer m, mG denotes the set of elements mg, $g \in G$, ${}_mG$ the set of elements $g \in G$ such that $mg = 0$, and G_m the factor group G/mG.

on to its residue class mod 2 and embeds $(H_{n+2})_2$ in $H_{n+2}(2)$. Moreover, $\Delta^{-1}(0)=\mu H_{n+2}$, so that $H_{n+2}(2)=\mu H_{n+2}+\Delta^*{}_2H_{n+1}$. Finally there is given a homomorphism $\gamma: H_{n+2}(2)\rightarrow(H_n)_2$.

Such a homology system is provided by the complex K. The groups $H_n, H_{n+1}, H_{n+2}, H_{n+2}(2)$ are interpreted as $H_n(K), H_{n+1}(K)$, $H_{n+2}(K)$ and $H_{n+2}(2)$ (defined in K as $\delta_{n+2}^{-1}(2Z_{n+1}(K))$, reduced mod 2). Then $(H_{n+2}(K))_2, =(Z_{n+2}(K))_2$, is obviously embedded in $H_{n+2}(2)$, so that μ is defined. The homomorphism γ is just $\nu(2)$, and we define $\Delta: H_{n+2}(2)\rightarrow H_{n+1}(K)$ as follows. Let x' be a chain in the mod 2 class $x\in H_{n+2}(2)$. Then $\frac{1}{2}\delta_{n+2}x'$ is an $(n+1)$-cycle whose homology class is Δx. Let $t_1,\ldots,t_m$ generate T_e, the subgroup of $H_{n+1}(K)$ of even torsions, so that t_i is of even order τ_i, $i=1,\ldots,m$. Then ${}_2H_{n+1}(K)$ is generated by $\frac{\tau_1}{2}t_1,\ldots,\frac{\tau_m}{2}t_m$. Let t'_i be a cycle in the class t_i, let u'_i be a $(n+2)$-chain such that $\delta_{n+2}u'_i=\tau_i t'_i$ and let u_i be the class, in $H_{n+2}(2)$, of u'_i. Then a right inverse $\Delta^*: {}_2H_{n+1}(K)\rightarrow H_{n+2}(2)$ is defined by $\Delta^*\frac{\tau_i}{2}t_i=u_i$, $i=1,\ldots,m$. It should be noted that Δ^* depends on the choice of the u'_i, so that it is only determined modulo an arbitrary homomorphism of ${}_2H_{n+1}(K)$ into $\mu H_{n+2}(K)$. It is then an easy matter to verify that, in K, $H_{n+2}(2)=\mu H_{n+2}(K)+\Delta^*{}_2H_{n+1}(K)$.

We now prescribe an arbitrary homology system and then construct a 'reduced' complex whose homology system is isomorphic to the given homology system. We will use the same symbols for the groups and homomorphisms of the abstract system as for the corresponding elements of the homology system of the complex, preferring γ to $\nu(2)$ for the homomorphism $H_{n+2}(2)\rightarrow(H_n)_2$. Our notation will then be in line with that of the author's paper.†

Let $H_n=(a_1,\ldots,a_m)$, where $\sigma_1a_1=\ldots=\sigma_ta_t=0$ and σ_i is odd if and only if $1\leqslant i\leqslant h\leqslant t\leqslant m$. In fact, we will take $\sigma_1,\ldots,\sigma_t$ to be torsion coefficients. Let $H_{n+1}=(b_1,\ldots,b_l)$, where

$$\tau_1b_1=\tau_2b_2=\ldots=\tau_ub_u=0$$

and τ_i is odd if and only if $1\leqslant i\leqslant k\leqslant u\leqslant l$. Again, we will take

† See footnote, p. 114.

$\tau_1, \ldots, \tau_u$ to be torsion coefficients. Let $H_{n+2} = (c_1, \ldots, c_p)$, being free Abelian. Then $(H_n)_2 = (\bar{a}_{h+1}, \ldots, \bar{a}_m)$, where $\bar{a}_i$ is the residue class mod 2 of a_i; similarly $(H_{n+2})_2 = (\bar{c}_1, \ldots, \bar{c}_p)$. The group ${}_2H_{n+1}$ is generated by $\frac{\tau_{k+1}}{2} b_{k+1}, \ldots, \frac{\tau_u}{2} b_u$, so that

$$H_{n+2}(2) = (\bar{c}_1, \ldots, \bar{c}_p, \bar{c}_{p+1}, \ldots, \bar{c}_{p+u-k}),$$

where $$\bar{c}_{p+i} = \Delta^* \frac{\tau_{k+i}}{2} b_{k+i}, \quad i = 1, \ldots, u-k.$$

Let $\gamma : H_{n+2}(2) \to (H_n)_2$ be given by

$$\gamma \bar{c}_i = \sum_{j=h+1}^{m} \gamma_{ij} \bar{a}_j, \quad i = 1, \ldots, p+u-k, \tag{3·1}$$

where the coefficients γ_{ij} are 0 or 1.

We now construct a cell-complex K as follows:

$K^0 \quad = \ldots = K^{n-1} = e^0$, a single point;

$K^n \quad = e^0 \cup e_1^n \cup \ldots \cup e_m^n$, so that e^0 closes e_i^n to an n-sphere S_i^n, $i = 1, \ldots, m$;

$K^{n+1} = K^n \cup e_1^{n+1} \cup \ldots \cup e_{t+l}^{n+1}$, where e_i^{n+1} is attached to K^n by a map $g_i : \dot{E}_i^{n+1} \to S_i^n$ of degree σ_i, $i = 1, \ldots, t$, and e^0 closes e_{t+j}^{n+1} to an $(n+1)$-sphere S_j^{n+1}, $j = 1, \ldots, l$;

$K^{n+2} = K^{n+1} \cup e_1^{n+2} \cup \ldots \cup e_{p+u}^{n+2}$, where e_i^{n+2} is attached to K^{n+1} by a map $f_i : \dot{E}_i^{n+2} \to K^n$, which is essential over those spheres S_j^n for which $\gamma_{ij} = 1, i = 1, \ldots, p$; e_{p+i}^{n+2} is attached to K^{n+1} by a map $f_{p+i} : \dot{E}_{p+i}^{n+2} \to K^{n+1}$ which is of degree τ_{k+i} over S_{k+i}^{n+1} and essential over those spheres S_j^n for which $\gamma_{p+i,j} = 1, i = 1, \ldots, u-k$; and $e_{p+u-k+i}^{n+2}$ is attached to K^{n+1} by a map $f_{p+u-k+i} : \dot{E}_{p+u-k+i}^{n+2} \to K^{n+1}$ which is of degree τ_i over S_i^{n+1}, $i = 1, \ldots, k$.

To clarify our description of the way in which the $(n+2)$-cells are attached, it is necessary to evaluate $\pi_{n+1}(K^{n+1})$. We may do this directly from the cell-decomposition of K^{n+1}. Alternatively, we may argue† from the exact sequence

$$H_{n+2}(K^{n+1}) \xrightarrow{\nu} \Gamma_{n+1}(K^{n+1}) \xrightarrow{\lambda} \pi_{n+1}(K^{n+1}) \xrightarrow{\mu} H_{n+1}(K^{n+1})$$

that μ is on to $H_{n+1}(K^{n+1})$, since $\pi_r(K^{n+1}) = \pi_r(K) = 0$, $r < n$. Also

† This argument is more satisfactory because it applies to any CW-complex whose first $(n-1)$ homotopy groups vanish.

$H_{n+2}(K^{n+1})=0$, so that the kernel of μ is the isomorphic image of $\Gamma_{n+1}(K^{n+1})$, and, by definition, $\Gamma_{n+1}(K^{n+1})$ is just $\Gamma_{n+1}(K)$ which is isomorphic to $(H_n)_2$. Thus

$$\pi_{n+1}(K^{n+1})-\lambda\Gamma_{n+1}(K^{n+1})\approx H_{n+1}(K^{n+1}),$$

and
$$\lambda\Gamma_{n+1}(K^{n+1})\approx (H_n)_2;$$

but since $H_{n+1}(K^{n+1})$ is free Abelian, it follows that

$$\pi_{n+1}(K^{n+1})\approx H_{n+1}(K^{n+1})+(H_n)_2. \tag{3·2}$$

Moreover, it is clear that $H_{n+1}(K^{n+1})$ is embedded in $\pi_{n+1}(K^{n+1})$ as the direct sum $\pi_{n+1}(S_1^{n+1})+\ldots+\pi_{n+1}(S_l^{n+1})$, and $(H_n)_2$ is embedded in $\pi_{n+1}(K^{n+1})$ as the direct sum

$$\pi_{n+1}(S^n_{h+1}\cup e^{n+1}_{h+1})+\ldots$$
$$+\pi_{n+1}(S^n_t\cup e^{n+1}_t)+\pi_{n+1}(S^n_{t+1})+\ldots+\pi_{n+1}(S^n_m).$$

The groups $\pi_{n+1}(S^n_i\cup e^{n+1}_i)$, $i=h+1,\ldots,t$, are just the isomorphic images under injection of $\pi_{n+1}(S^n_i)$.

It is a straightforward matter to verify that the cell-complex K realizes the given abstract homology system. We will just carry through the verification that $\nu(2)$ coincides with the given homomorphism γ.

Now $C_{n+2}(K)$ is generated by $(c_1,\ldots,c_{p+u})$, where c_i corresponds to the $(n+2)$-cell e_i^{n+2} and the subgroup $\delta_{n+2}^{-1}(2Z_{n+1}(K))$ is generated by $(c_1,\ldots,c_{p+u-k})$; thus $H_{n+2}(2)=(\bar{c}_1,\ldots,\bar{c}_{p+u-k})$. We recall that $\nu(2)(\bar{c}_i)$ is the element $d_{n+2}(c_i)-2y_i$, where $d_{n+2}:\pi_{n+2}(K^{n+2},K^{n+1})\to\pi_{n+1}(K^{n+1})$ is the homotopy boundary, and y_i is an element in $\pi_{n+1}(K^{n+1})$ such that

$$2j_{n+1}(y_i),\in\pi_{n+1}(K^{n+1},K^n),=\delta_{n+2}(c_i).$$

Identifying $H_{n+1}(K^{n+1})$, by means of (3·2), with a subgroup of $\pi_{n+1}(K^{n+1})$, we have $\nu(2)(\bar{c}_i)=d_{n+2}(c_i)-\delta_{n+2}(c_i)$. Now

$$d_{n+2}(c_i)=\sum_{j=h+1}^{m}\gamma_{ij}\bar{a}_j,\quad i=1,\ldots,p;\quad \delta_{n+2}(c_i)=0,\quad i=1,\ldots,p,$$

$$d_{n+2}(c_{p+i})=\tau_{k+i}b^*_{k+i}+\sum_{j=h+1}^{m}\gamma_{p+i,j}\bar{a}_j,\quad i=1,\ldots,u-k;$$

$$\delta_{n+2}(c_{p+i})=\tau_{k+i}b^*_{k+i},\quad i=1,\ldots,u-k,$$

when we identify $(H_n)_2$ with a subgroup of $\pi_{n+1}(K^{n+1})$ by means of (3·2), and where $H_{n+1}(K^{n+1}) = (b_1^*, \ldots, b_l^*)$, b_i^* corresponding to S_i^{n+1}, $i = 1, \ldots, l$. Then

$$\nu(2)(\bar{c}_i) = \sum_{j=h+1}^{m} \gamma_{ij}\bar{a}_j, \quad i = 1, \ldots, p+u-k, \quad = \gamma(\bar{c}_i),$$

as was to be proved.

We may now compute $\pi_{n+1}(K)$ by considering the relations introduced into $\pi_{n+1}(K^{n+1})$ by attaching the $(p+u)$ $(n+2)$-cells $e_1^{n+2}, \ldots, e_{p+u}^{n+2}$. With the identifications made earlier we have

$$\pi_{n+1}(K^{n+1}) = (b_1^*, \ldots, b_l^*) + (\bar{a}_{h+1}, \ldots, \bar{a}_m), \tag{3·3}$$

and it is not difficult to see that the relations introduced are

$$\left.\begin{aligned} &\gamma\bar{c}_i = \Sigma\gamma_{ij}\bar{a}_j = 0, \quad i = 1, \ldots, p, \\ \tau_{k+i}b_{k+i}^* + \gamma\bar{c}_{p+i} = \tau_{k+i}&b_{k+i}^* + \Sigma\gamma_{p+i,j}\bar{a}_j = 0, \quad i = 1, \ldots, u-k, \\ &\tau_i b_i^* = 0, \quad i = 1, \ldots, k. \end{aligned}\right\} \tag{3·4}$$

This leads at once to a formulation of the value of $\pi_{n+1}(K)$ almost identical in form with that given in § 2. We refer to the author's previously quoted paper for the details of the derivation of the following theorem† which may be regarded as summing up the contents of § 2 and the present section, in so far as they relate to the calculation of $\pi_{n+1}(K)$.

THEOREM 3·5. *If K is a finite cell-complex of dimension $\leqslant n+2$, $n > 2$, such that $\pi_r(K) = 0$, $r = 1, \ldots, n-1$, then $\pi_{n+1}(K)$ is an extension of H_{n+1} by $(H_n)_2 - \gamma\mu H_{n+2}$. If H_{n+1} is given as $B + T_o + T_e$, where B is free Abelian, T_o is a group of odd torsions, and T_e is a group of even torsions, then $\pi_{n+1}(K) \approx B + T_o + E$, where E is an extension of T_e by $(H_n)_2 - \gamma\mu H_{n+2}$. Let T_e be generated by $b_{k+1}, \ldots, b_u$ of orders $\tau_{k+1}, \ldots, \tau_u$. Then representatives of $b_{k+1}, \ldots, b_u$ may be chosen in E, and called $b_{k+1}^*, \ldots, b_u^*$, such that $\tau_{k+i}b_{k+i}^* = j\gamma\Delta^{-1}\frac{1}{2}\tau_{k+i}b_{k+i}$, where j is the projection*

$$(H_n)_2 \to (H_n)_2 - \gamma\mu H_{n+2}.$$

† This is theorem 4·1 of the paper quoted in footnote, p. 114.

J. H. C. Whitehead has shown that the homology system determines the homotopy type of a cell-complex† having the properties of K in theorem 3·5. Thus we are led to consider the problem of calculating the higher homotopy groups of K. Such calculations depend, of course, on the homotopy groups of spheres. Using the fact that $\pi_{n+2}(S^n)$ is cyclic of order 2, $n \geqslant 2$, we may calculate $\pi_{n+2}(K)$, but it is, in general, rather complicated.

Let K be $(n+2)$-dimensional. Then, from the sequence

$$H_{n+3}(K) \xrightarrow{\nu_{n+3}} \Gamma_{n+2}(K) \xrightarrow{\lambda_{n+2}} \pi_{n+2}(K) \xrightarrow{\mu_{n+2}} H_{n+2}(K) \xrightarrow{\nu_{n+2}} \Gamma_{n+1}(K),$$

we have $H_{n+3}(K)=0$, so that λ_{n+2} is isomorphic and μ_{n+2} is a homomorphism of $\pi_{n+2}(K)$ on to $\nu_{n+2}^{-1}(0)$ with kernel isomorphic to $\Gamma_{n+2}(K)$. Moreover, $\nu_{n+2}^{-1}(0)$, being a subgroup of the free Abelian group $H_{n+2}(K)$, is itself free Abelian, so that

$$\pi_{n+2}(K) \approx \Gamma_{n+2}(K) + \nu_{n+2}^{-1}(0), \tag{3·6}$$

and the calculation of $\pi_{n+2}(K)$ is referred back to that of $\Gamma_{n+2}(K)$. Now suppose $n>2$ and $\pi_r(K)=0$, $r=1, \ldots, n-1$. Then $\Gamma_{n+1}(K)$ is a free mod 2 module so that $\nu_{n+2}(2H_{n+2}(K))=0$. Since $H_{n+2}(K)$ is free Abelian, it follows that $2H_{n+2}(K) \approx H_{n+2}(K)$ and, since $\nu_{n+2}^{-1}(0)$ is a group 'between' $H_{n+2}(K)$ and $2H_{n+2}(K)$, we also have $\nu_{n+2}^{-1}(0) \approx H_{n+2}(K)$. Thus, in this case,

$$\pi_{n+2}(K) \approx \Gamma_{n+2}(K) + H_{n+2}(K). \tag{3·7}$$

We now calculate $\Gamma_{n+2}(K)$ in a special case.

THEOREM 3·8. *If K is as in theorem* 3·5, *if $n>3$, and if $\pi_{n+1}(K)=0$, then $\Gamma_{n+2}(K) \approx {}_2H_n$.*

Since $\pi_{n+1}(K)=0$, it follows from theorem 3·5 that $H_{n+1}(K)=0$ and $\gamma\mu H_{n+2}=(H_n)_2$. We may thus choose a basis $\bar{c}_1, \ldots, \bar{c}_q$ for μH_{n+2}, such that

$$\gamma\bar{c}_i = \bar{a}_{h+i}, \quad i=1, \ldots, m-h$$
$$\gamma\bar{c}_j = 0, \qquad j=m-h+1, \ldots, q.$$

† He has also shown that the homotopy type of a (perhaps infinite) CW-complex of dimension $\leqslant n+2$, $n>2$, whose first $(n-1)$ homotopy groups vanish, is determined by the part of its 'exact sequence' beginning

$$H_{n+2}(K) \xrightarrow{\nu_{n+2}} \Gamma_{n+1}(K) \longrightarrow \ldots.$$

Since H_{n+2} is free Abelian, the basis $\bar{c}_1, \ldots, \bar{c}_q$ can be 'lifted' to a basis $c_1, \ldots, c_q$ for H_{n+2}. A reduced complex of the same homotopy type as K, which we may take to be K itself, will then have the following specifications:

$K^0 \ = \ldots = K^{n-1} = e^0$;

$K^n \ = e^0 \cup e_1^n \cup \ldots \cup e_m^n$, *and* $e^0 \cup e_i^n = S_i^n$, $i = 1, \ldots, m$;

$K^{n+1} = K^n \cup e_1^{n+1} \cup \ldots \cup e_t^{n+1}$, where e_i^{n+1} is attached to K^n by a map $g_i : \dot{E}_i^{n+1} \to S_i^n$ of degree σ_i, $i = 1, \ldots, t$, σ_i being odd if and only if $1 \leqslant i \leqslant h \leqslant t$;

$K^{n+2} = K^{n+1} \cup e_1^{n+2} \cup \ldots \cup e_q^{n+2}$, where e_i^{n+2} is attached to K^n by an essential map $f_i : \dot{E}_i^{n+2} \to S_{h+i}^n$, $i = 1, \ldots, m-h$, and $e^0 \cup e_j^{n+2} = S_j^{n+2}$, $j = m-h+1, \ldots, q$.

We now compute $\pi_{n+2}(K^{n+1})$. Now if $n > 3$, $n+2 < 2n-1$, so that,† by (4·3) of Chapter VI,

$$\pi_{n+2}(S_1^n \cup S_2^n) \approx \pi_{n+2}(S_1^n) + \pi_{n+2}(S_2^n), \quad n > 3, \tag{3·9}$$

and this result may clearly be extended to

$$\pi_{n+2}(K^n) \approx \pi_{n+2}(S_1^n) + \ldots + \pi_{n+2}(S_m^n), \tag{3·10}$$

each direct summand $\pi_{n+2}(S_i^n)$, $i = 1, \ldots, m$, being embedded isomorphically in $\pi_{n+2}(K^n)$ by injection. We will represent $\pi_{n+2}(K^n)$ as $(\alpha_1, \ldots, \alpha_m)$, where α_i corresponds to S_i^n, $i = 1, \ldots, m$.

Let $g_1 : \dot{E}_1^{n+1} \to S_1^n$ induce‡ $h_r : \pi_r(\dot{E}_1^{n+1}) \to \pi_r(K^n)$. Since σ_1 is odd, and since

$$\pi_{n+2}(S^n) = F\pi_{n+1}(S^{n-1}), \quad \pi_{n+1}(S^n) = F\pi_n(S^{n-1}),$$

it follows readily that $h_r : \pi_r(\dot{E}_1^{n+1}) \approx \pi_r(S_1^n)$, $r = n+1, n+2$. If g_1 is extended to a characteristic map

$$g_1' : E_1^{n+1}, \dot{E}_1^{n+1} \to K^n \cup e_1^{n+1}, K^n$$

and if g_1' induces

$$\bar{g}_r : \pi_r(E_1^{n+1}, \dot{E}_1^{n+1}) \to \pi_r(K^n \cup e_1^{n+1}, K^n),$$

then it follows from theorem 2·13 of Chapter VI that

$$\bar{g}_r : \pi_r(E_1^{n+1}, \dot{E}_1^{n+1}) \approx \pi_r(K^n \cup e_1^{n+1}, K^n), \quad r = n+2, n+3.$$

† It is at this point that the restriction to $n > 3$ is necessary. The case $n = 3$ is also treated in the author's paper, *Quart. J. Math.* (Oxford), (2), 2 (1951), 228–40, theorem 5·10.

‡ We regard $\pi_r(S_1^n)$ as embedded in $\pi_r(K^n)$ by injection; this part of the argument is, of course, void if K has no odd torsion.

Consider the diagram

$$\begin{array}{ccccc}
\pi_{n+3}(E_1^{n+1}, \dot{E}_1^{n+1}) & \xrightarrow{d'_{n+3}} & \pi_{n+2}(\dot{E}_1^{n+1}) & & \\
\downarrow \bar{g}_{n+3} & & \downarrow h_{n+2} & & \\
\pi_{n+3}(K^n \cup e_1^{n+1}, K^n) & \xrightarrow{d_{n+3}} & \pi_{n+2}(K^n) & \xrightarrow{i_{n+2}} & \pi_{n+2}(K^n \cup e_1^{n+1})
\end{array}$$

$$\begin{array}{ccccc}
 & & \pi_{n+2}(E_1^{n+1}, \dot{E}_1^{n+1}) & \xrightarrow{d'_{n+2}} & \pi_{n+2}(\dot{E}_1^{n+1}) \\
 & & \downarrow \bar{g}_{n+2} & & \downarrow h_{n+1} \\
\pi_{n+2}(K^n \cup e_1^{n+1}) & \xrightarrow{j_{n+2}} & \pi_{n+2}(K^n \cup e_1^{n+1}, K^n) & \xrightarrow{d_{n+2}} & \pi_{n+1}(K^n)
\end{array}$$

We take the meanings of the various homomorphisms to be, by now, self-evident, as also the fact that $d_{n+3}\bar{g}_{n+3} = h_{n+2}d'_{n+3}$, $d_{n+2}\bar{g}_{n+2} = h_{n+1}d'_{n+2}$.

Since $\bar{g}_{n+2}$ is an isomorphism and $d_{n+2}\bar{g}_{n+2} = h_{n+1}d'_{n+2}$, it follows that $\bar{g}_{n+2}d'^{-1}_{n+2}$ maps $h^{-1}_{n+1}(0)$ isomorphically on to $d^{-1}_{n+2}(0)$; but

$$d^{-1}_{n+2}(0) = j_{n+2}\pi_{n+2}(K^n \cup e_1^{n+1}) \approx \pi_{n+2}(K^n \cup e_1^{n+1}) - i_{n+2}\pi_{n+2}(K^n).$$

Also

$$\begin{aligned}
i^{-1}_{n+2}(0) &= d_{n+3}\pi_{n+3}(K^n \cup e_1^{n+1}, K^n) = d_{n+3}\bar{g}_{n+3}\pi_{n+3}(E_1^{n+1}, \dot{E}_1^{n+1}) \\
&= h_{n+2}d'_{n+3}\pi_{n+3}(E_1^{n+1}, \dot{E}_1^{n+1}) \doteq h_{n+2}\pi_{n+2}(\dot{E}_1^{n+1}).
\end{aligned}$$

We now use the fact that σ_1 is odd, whence $h^{-1}_{n+1}(0) = 0$, and

$$h_{n+2}\pi_{n+2}(\dot{E}_1^{n+1}) = \pi_{n+2}(S_1^n).$$

Thus $\pi_{n+2}(K^n \cup e_1^{n+1}) - i_{n+2}\pi_{n+2}(K^n) = 0$, so that, if we represent $\pi_{n+2}(K^n)$ as $(\alpha_1, \ldots, \alpha_m)$, then $\pi_{n+2}(K^n \cup e_1^{n+1})$ is obtained by suppressing the generator α_1.

Clearly we may proceed in this way till we have attached the first h $(n+1)$-cells to K^n, forming, say, K_0^{n+1}, and obtain

$$\pi_{n+2}(K_0^{n+1}) = (\alpha_{h+1}, \ldots, \alpha_m). \tag{3·11}$$

Consider the attachment of e^{n+1}_{h+1} to K_0^{n+1}. A precisely similar argument to the one above† shows that $h^{-1}_{n+1}(0)$ is isomorphic to

$$\pi_{n+2}(K_0^{n+1} \cup e^{n+1}_{h+1}) - i_{n+2}\pi_{n+2}(K_0^{n+1})$$

and $$i^{-1}_{n+2}(0) = h_{n+2}\pi_{n+2}(\dot{E}^{n+1}_{h+1}),$$

† We are, in effect, using theorem 8 of J. H. C. Whitehead's paper 'On $\pi_r(V_{n,m})$ and sphere-bundles', *Proc. Lond. Math. Soc.* (2), 48 (1944), 243–91.

where $$h_r : \pi_r(\dot{E}^{n+1}_{h+1}) \to \pi_r(K^{n+1}_0)$$

is induced by† g_{h+1}. Since σ_{h+1} is even,

$$h_{n+2}\pi_{n+2}(\dot{E}^{n+1}_{h+1}) = h_{n+1}\pi_{n+1}(\dot{E}^{n+1}_{h+1}) = 0,$$

so that $\pi_{n+2}(K^{n+1}_0 \cup e^{n+1}_{h+1})$ is an extension of a cyclic group of order 2 by $\pi_{n+2}(K^{n+1}_0)$. Proceeding in this way for each of the remaining $(t-h)$ $(n+1)$-cells, we find that $\pi_{n+2}(K^{n+1})$ is an extension of a free mod 2 module on $(t-h)$ generators by $\pi_{n+2}(K^{n+1}_0)$. If B, $= (\beta_{h+1}, \ldots, \beta_t)$, is the free mod 2 module on $(t-h)$ generators and $A = \pi_{n+2}(K^{n+1}_0) = (\alpha_{h+1}, \ldots, \alpha_m)$, then $\pi_{n+2}(K^{n+1})$ is an extension of B by A. It should be noted that, in fact, $\pi_{n+2}(K^{n+1})$ is the direct sum

$$\pi_{n+2}(S^n_{h+1} \cup e^{n+1}_{h+1}) + \ldots + \pi_{n+2}(S^n_t \cup e^{n+1}_t) + \pi_{n+2}(S^n_{t+1}) + \ldots + \pi_{n+2}(S^n_m)$$

and $\pi_{n+2}(S^n_i \cup e^{n+1}_i)$ is an extension of (β_i) by (α_i), $i = h+1, \ldots, t$. We do not determine the precise nature of the group extension,‡ because we will not require it here.

Finally, we compute $\Gamma_{n+2}(K)$. We have to study the effect on $\pi_{n+2}(K^{n+1})$ of attaching the $(n+2)$-cells of K. This, however, presents little difficulty. For, as may readily be seen, the attachment of each $(n+2)$-cell e^{n+2}_i, $i = 1, \ldots, m-h$, simply 'kills' the generator α_{h+i}, and, of course, the attachment of the $(n+2)$-spheres S^{n+2}_j, $j = m-h+1, \ldots, q$, does not affect $\Gamma_{n+2}(K)$. Thus we obtain $\Gamma_{n+2}(K)$ from $\pi_{n+2}(K^{n+1})$ by 'killing' the subgroup A, whence $\Gamma_{n+2}(K) \approx B$. On the other hand, H_n is the direct sum of a free Abelian group, a group of odd order, and $(t-h)$ cyclic groups of even order, so that $B \approx {}_2H_n$. Thus the theorem is proved.

4. Normal complex of S. C. Chang. In his paper 'Homotopy invariants and continuous mappings' (*Proc. Roy. Soc.* A, 202 (1950), 253–63), S. C. Chang introduced new numerical invariants, the *secondary torsions*, which, together with the Betti

† We regard $\pi_r(S^n_{h+1})$ as embedded in $\pi_r(K^{n+1}_0)$ by injection.

‡ Actually, if e^{n+1} is attached to S^n by a map of even degree σ, then $\pi_{n+2}(S^n \cup e^{n+1}) = Z_4$ if 4 does not divide σ, $\pi_{n+2}(S^n \cup e^{n+1}) = Z_2 + Z_2$ if 4 divides σ.

numbers and torsions of a finite cell-complex† K, of dimension $\leqslant n+2$, $n>2$, with $\pi_r(K)=0$, $1\leqslant r<n$, determine the homotopy type of K. These secondary torsions might be calculated from a simplicial decomposition of K, and enabled Chang to describe the *normal form* for a complex of the same homotopy type as K. We will not introduce the secondary torsions here, but will define a *normal* cell-complex.

A *normal* cell-complex consists of a finite number of *elementary* complexes joined together at a single common point. Chang divided the elementary complexes into *seven* types, but we will make a few further subdivisions and list below the *eleven* types of elementary complex.

Type 1. S^n.

Type 2. S^{n+1}.

Type 3. S^{n+2}.

Type 4. $S^n \cup e^{n+1}$, where e^{n+1} is attached to S^n by a map of degree 2^q, $q>0$.

Type 5. $S^n \cup e^{n+1}$, where e^{n+1} is attached to S^n by a map of degree p^q, $q>0$, p an odd prime.

Type 6. $S^n \cup e^{n+2}$, where e^{n+2} is attached to S^n by an essential map $\dot{E}^{n+2}\to S^n$.

Type 7. $S^n \cup e^{n+1} \cup e^{n+2}$, where $S^n \cup e^{n+1}$ is of type 4, $S^n \cup e^{n+2}$ of type 6.

Type 8. $S^{n+1} \cup e^{n+2}$, where e^{n+2} is attached to S^{n+1} by a map of degree $2^{q'}$, $q'>0$.

Type 9. $S^{n+1} \cup e^{n+2}$, where e^{n+2} is attached to S^{n+1} by a map of degree $p^{q'}$, $q'>0$, p an odd prime.

Type 10. $S^n \cup S^{n+1} \cup e^{n+2}$, where e^{n+2} is attached to $S^n \cup S^{n+1}$ by a map which is essential over S^n and of degree $2^{q'}$, $q'>0$, over S^{n+1}.

Type 11. $S^n \cup e^{n+1} \cup S^{n+1} \cup e^{n+2}$, where $S^n \cup e^{n+1}$ is of type 4, $S^n \cup S^{n+1} \cup e^{n+2}$ of type 10.

† The secondary torsions may, in fact, be defined for any finite complex.

To calculate $\pi_{n+1}(K)$, it is now only necessary to replace K by a normal complex of the same homotopy type, and then to take the direct sum of the $(n+1)$st homotopy groups of the constituent elementary complexes. The first operation requires the secondary torsions (or, what is equivalent, the reduction of γ to normal form, γ being the homomorphism $H_{n+2}(2) \to (H_n)_2$) and the second operation may be based on the following table, where K_i represents a standard complex of type i, and Z_m is a cyclic group of order m:

i	=	1	2	3	4	5	6	7	8	9	10	11
$\pi_{n+1}(K_i)$	=	Z_2	Z_∞	0	Z_2	0	0	0	$Z_{2^{q'}}$	$Z_{p^{q'}}$	$Z_{2^{q'+1}}$	$Z_{2^{q'+1}}$

These results are, of course, all special cases of theorem 3·5. The fact that $\pi_{n+1}(K)$ is then just the direct sum of the $(n+1)$st homotopy groups of the constituent elementary complexes requires proof, but it is, actually, an easy consequence of (4·3) of Chapter VI and the Whitehead suspension theorems. In general, in computing $\pi_r(K)$, for arbitrary r, the direct sum of the rth homotopy groups of the constituent elementary complexes appears as a direct summand in $\pi_r(K)$, the remaining direct summand being called by the author the *cross-term*. Thus, for example, if $r = n+2$, the cross-term is zero if $n > 3$,† but is non-zero if $n = 3$.

Theorem 3·8 follows readily by the use of elementary complexes. We re-prove it as

THEOREM 4·1. *If K is a finite cell-complex of dimension $\leqslant n+2$, $n > 3$, such that $\pi_r(K) = 0$, $r = 1, \ldots, n-1, n+1$, then*

$$\pi_{n+2}(K) \approx H_{n+2}(K) + {}_2H_n(K).$$

Let us assume from the outset that K is in normal form. Then it can only contain constituent elementary complexes of types 3, 5, 6 and 7. We will not stop to prove here that $\pi_{n+2}(K)$ is, in fact, the direct sum of the $(n+2)$nd homotopy groups of its constituents, since this is an immediate consequence of theorem 4·2, but will content ourselves with proving that the stated direct sum is isomorphic to $H_{n+2}(K) + {}_2H_n(K)$.

† See theorem 4·2.

Suppose K is composed of k_i elementary complexes of type i, $i=3, 5, 6, 7$. Then $H_{n+2}(K)$ is free Abelian of rank $(k_3+k_6+k_7)$ and ${}_2H_n$ is a free mod 2 module of rank k_7.

Let K_i be an elementary complex of type i, $i=1, \dots, 11$.

We now prove

(i) $\pi_{n+2}(K_5)=0$, (ii) $\pi_{n+2}(K_6)=Z_\infty$, (iii) $\pi_{n+2}(K_7)=Z_2+Z_\infty$.

Proof of (i). It is only necessary to simplify the argument in the proof of theorem 3·8, between (3·10) and (3·11).

Proof of (ii). We could again argue as in theorem 3·8. If the attaching map for e^{n+2} induces $h_r: \pi_r(\dot{E}^{n+2}) \to \pi_r(S^n)$, then $h_{n+1}^{-1}(0) \approx \pi_{n+2}(K_6) - i\pi_{n+2}(S^n)$ and $i^{-1}(0)=h_{n+2}\pi_{n+2}(\dot{E}^{n+2})$. Now

$$h_{n+1}^{-1}(0)=2\pi_{n+1}(S^{n+1})=Z_\infty \text{ and } h_{n+2}\pi_{n+2}(\dot{E}^{n+2})=\pi_{n+2}(S^n),$$

whence $i\pi_{n+2}(S^n)=0$ and $\pi_{n+2}(K_6)=Z_\infty$.

Proof of (iii). If the attaching map for e^{n+2} induces

$$h_r: \pi_r(\dot{E}^{n+2}) \to \pi_r(S^n \cup e^{n+1}),$$

then $$h_{n+1}^{-1}(0) \approx \pi_{n+2}(K_7) - i\pi_{n+2}(S^n \cup e^{n+1})$$

and $$i^{-1}(0)=h_{n+2}\pi_{n+2}(\dot{E}^{n+2}).$$

Now it was proved, in the course of the proof of theorem 3·8, that $\pi_{n+2}(S^n \cup e^{n+1})$ contains $\pi_{n+2}(S^n)$ as a subgroup and the factor group is Z_2. Since $h_{n+2}\pi_{n+2}(\dot{E}^{n+2})=\pi_{n+2}(S^n)$, it follows that $i\pi_{n+2}(S^n \cup e^{n+1})=Z_2$. Also $h_{n+1}^{-1}(0)=2\pi_{n+1}(\dot{E}^{n+2})=Z_\infty$, so that $\pi_{n+2}(K_7)$ is an extension (and, therefore, a trivial extension) of Z_∞ by Z_2.

Finally, we prove a theorem which establishes, as a special case, the result that $\pi_{n+2}(K)$ is just the direct sum of the $(n+2)$nd homotopy groups of its constituents. This, together with (i), (ii), (iii) above, establishes theorem 4·1.

THEOREM 4·2. *Let $K_{(r)}$, $r=1, \dots, m$, be a finite cell-complex such that $K_{(r)}^{n-1}=e_{(r)}^0$, and let K be the union of the $K_{(r)}$ with the $e_{(r)}^0$ identified with a single vertex e^0. Then $\pi_s(K)=\Sigma\pi_s(K_{(r)})$, provided $1 < s < 2n-1$.*

Let $Y_1, \dots, Y_m$ be m arcwise-connected spaces and Y the union of the Y_i, $i=1, \dots, m$, with a single point on each identified. Then by a natural extension of the argument of theorem 6·2, Chapter IV, we have

$$\pi_s(Y) = \pi_s(Y_1) + \ldots + \pi_s(Y_m) + d\pi_{s+1}(Y_1 \times \ldots \times Y_m, Y), \quad s \geqslant 2, \quad (4\cdot3)$$

where $\pi_s(Y_i)$, $i = 1, \ldots, m$, is embedded in $\pi_s(Y)$ by injection and d is the usual homotopy boundary homomorphism, which, in this case, is isomorphic. The theorem will then be proved if we can show that

$$\pi_{s+1}(K_{(1)} \times \ldots \times K_{(m)}, K) = 0, \quad \text{if} \quad s+1 < 2n.$$

Now consider the cell decomposition of $K_{(1)} \times \ldots \times K_{(m)}$. It consists of K together with certain additional cells whose dimensionality must be at least $2n$. Let

$$f: I^{s+1}, \dot{I}^{s+1} \to K_{(1)} \times \ldots \times K_{(m)}, K$$

represent α, an arbitrary element of $\pi_{s+1}(K_{(1)} \times \ldots \times K_{(m)}, K)$. We may represent I^{s+1} as the union of the cells $e^0 \cup e^s \cup e^{s+1}$ where $\dot{I}^{s+1} = e^0 \cup e^s$. Then, by theorem 1·8 of Chapter VII, we may deform $f \mid \dot{I}^{s+1}$ so that the resulting map of $\dot{I}^{s+1}$ into K is cellular, and, of course, so that the image of $\dot{I}^{s+1}$ remains in K throughout the deformation. This deformation may be extended to a deformation of f, so that we may assume from the outset that $f \mid \dot{I}^{s+1}$ is cellular. Then we may deform f into a cellular map, keeping $f \mid \dot{I}^{s+1}$ fixed. If f' is the result of the deformation, then f' still represents α and

$$f'(I^{s+1}) \subset (K_{(1)} \times \ldots \times K_{(m)})^{s+1} \subset (K_{(1)} \times \ldots \times K_{(m)})^{2n-1},$$

since $s+1 < 2n$. However, since no cell of dimension less than $2n$ is attached to K to form $K_{(1)} \times \ldots \times K_{(m)}$, it follows that $(K_{(1)} \times \ldots \times K_{(m)})^{2n-1} \subset K$, so that $f'(I^{s+1}) \subset K$ and $\alpha = 0$. This establishes the theorem.

Theorem 4·2 admits generalization. Thus the $K_{(r)}$ may be replaced by CW-complexes whose first $(n-1)$ homotopy groups vanish. More generally, if $K_{(r)}$ is a CW-complex whose first $(n_r - 1)$ homotopy groups vanish, $n_r \geqslant 2$, then $\pi_s(K) = \Sigma \pi_s(K_{(r)})$, provided $1 < s < N + N' - 1$, where N, N' are the two smallest numbers† in the set $\{n_1, \ldots, n_m\}$. We may allow $m = \infty$ provided we give K the weak topology. J. H. C. Whitehead has shown

† We can, of course, have $N = N'$.

that, if X is an arcwise-connected topological space whose first $(m-1)$ homotopy groups vanish, then

$$\pi_s(X \cup S^n) = \pi_s(X) + \pi_s(S^n),$$

provided $1 < s < m+n-1$, and this theorem may readily be extended to that obtained by replacing S^n by a space Y whose first $(n-1)$ homotopy groups vanish.

5. Appendix.

Table of homotopy groups of elementary complexes†

Type	π_{n+1}, $n \geqslant 3$	π_{n+2}, $n \geqslant 3$	π_{n+3}, $n \geqslant 5$	π_7, $n=4$	π_6, $n=3$
1	Z_2	Z_2	Z_{24}	$Z_\infty + Z_{12}$	Z_{12}
2	Z_∞	Z_2	Z_2	Z_2	Z_2
3	0	Z_∞	Z_2	Z_2	Z_2
4	Z_2	$\left.\begin{matrix} Z_4, q=1 \\ Z_2+Z_2, q>1 \end{matrix}\right\}$	$Z_{(2^q, 24)} + Z_2$	$Z_{2^{q+1}} + Z_{(2^{q-1}, 12)}$	?
5	0	0	$Z_{(p^q, 24)}$	$Z_{p^q} + Z_{(p^q, 12)}$	?
6	0	Z_∞	Z_{12}	$Z_\infty + Z_6$	Z_6
7	0	$Z_2 + Z_\infty$	$Z_{(2^q, 12)} + Z_2$	$Z_{2^{q+1}} + Z_{(2^{q-1}, 6)} + Z_2$	?
8	$Z_{2^{q'}}$	Z_2	$\left.\begin{matrix} Z_4, q'=1 \\ Z_2+Z_2, q'>1 \end{matrix}\right\}$	$\left.\begin{matrix} Z_4, q'=1 \\ Z_2+Z_2, q'>1 \end{matrix}\right\}$	$\left.\begin{matrix} Z_4, q'=1 \\ Z_2+Z_2, q'>1 \end{matrix}\right\}$
9	$Z_{p^{q'}}$	0	0	0	0
10	$Z_{2^{q'+1}}$	Z_2	$Z_{12} + Z_2$	$Z_\infty + Z_6 + Z_2$	$Z_6 + Z_2 + Z_{2^{q'}}$
11	$Z_{2^{q'+1}}$	$Z_2 + Z_2$	$Z_{(2^q, 12)} + Z_2 + Z_2$	$Z_{2^{q+1}} + Z_{(2^{q-1}, 6)} + Z_2 + Z_2$	?

[Z_r = cyclic group of order r; (s, t) = greatest common divisor of s, t; + denotes, as usual, the direct sum.]

[The groups listed as unknown in this table have been computed by M. G. Barratt.]‡

† The result for $\pi_{n+2}(K)$ where K is elementary of type 4 is due to M. G. Barratt and G. F. Paechter if $q=1$, and to the author if $q>1$.

J.-P. Serre has recently calculated $\pi_{n+4}(S^n)$ and $\pi_{n+5}(S^n)$ as follows:

$\pi_6(S^2) = Z_{12}$, $\pi_7(S^3) = Z_2$, $\pi_8(S^4) = Z_2 + Z_2$, $\pi_9(S^5) = Z_2$, $\pi_{n+4}(S^n) = 0$, $n \geqslant 6$;

$\pi_7(S^2) = Z_2$, $\pi_8(S^3) = Z_2$, $\pi_9(S^4) = Z_2 + Z_2$, $\pi_{10}(S^5) = Z_2$, $\pi_{11}(S^6) = Z_\infty$, $\pi_{n+5}(S^n) = 0$, $n \geqslant 7$.

The generators of these groups are known.

‡ Note added in proof.

BIBLIOGRAPHY

(The books and papers listed below are not supposed to exhaust the literature on homotopy theory, but rather to be related to the subjects discussed in the text. Also, many of the results contained in the early papers on homotopy theory have been gathered together in the books by Lefschetz and Steenrod, so that a reference to the appropriate book becomes more convenient than a reference to the original paper. The numbers in square brackets after each item refer to the chapters of the text to which the contents of the item are particularly relevant.)

BOOKS

LEFSCHETZ, S. *Introduction to Topology*, no. 11, Princeton Mathematical Series, Princeton University Press, 1949. [3] (This work also serves as a general background to the text.)

NEWMAN, M. H. A. *Topology of Plane Sets*, 2nd ed., C.U.P., 1951. [1] (For an introduction to the notions of homotopy and deformation.)

SEIFERT, H. and THRELFALL, W. *Lehrbuch der Topologie*, Teubner, Leipzig, 1934. [6] (This is a standard introduction to combinatorial topology. The treatment of the fundamental group and covering complexes will be found particularly useful.)

STEENROD, N. E. *Topology of Fibre Bundles*, no. 14, Princeton Mathematical Series, Princeton University Press, 1951. [2, 5, 6]

PAPERS

BARRATT, M. G. and PAECHTER, G. F. Note on $\pi_r(V_{n,m})$. *Proc. Nat. Acad. Sci., Wash.*, 38 (1952), 119–21. [5, 6, 8]

BLAKERS, A. L. and MASSEY, W. S. The homotopy groups of a triad. I. *Ann. Math.* 53 (1951), 161–205. [2, 4]

BLAKERS, A. L. and MASSEY, W. S. The homotopy groups of a triad. II. *Ann. Math.* 55 (1952), 192–201. [2]

CHANG, S. C. Homotopy invariants and continuous mappings. *Proc. Roy. Soc.* A, 202 (1950), 253–63. [8]

CHANG, S. C. Some suspension theorems. *Quart. J. Math.* (Oxford), (2), 1 (1950), 310–17. [6]

COBBE, ANNE. Some algebraic properties of crossed modules. *Quart. J. Math.* (Oxford), (2), 2 (1951), 269–85. [4]

COCKCROFT, W. H. The word problem in a group extension. *Quart. J. Math.* (Oxford), (2), 2 (1951), 123–34. [4]

COCKCROFT, W. H. Note on a theorem by J. H. C. Whitehead. *Quart. J. Math.* (Oxford), (2), 2 (1951), 159–60. [4]

ECKMANN, B. Zur Homotopietheorie gefaserter Räume. *Comm. Math. Helv.* 14 (1942), 141–92. [5]

ECKMANN, B. Espaces Fibrés et Homotopie. *Colloque de Topologie* (Espaces Fibrés), Georges Thone, Liège (1951), pp. 83–99. [5]

EILENBERG, S. and MACLANE, S. Relations between homology and homotopy groups of spaces. *Ann. Math.* 46 (1945), 480–509. [4, 5]

EILENBERG, S. and MACLANE, S. Relations between homology and homotopy groups of spaces. II. *Ann. Math.* 51 (1950), 514–33. [4, 5]

EILENBERG, S. and STEENROD, N. E. Axiomatic approach to homology theory. *Proc. Nat. Acad. Sci., Wash.*, 31 (1945), 117–20. [4, 6]

FOX, R. H. On topologies for function spaces. *Bull. Amer. Math. Soc.* 51 (1945), 429–32. [5]

FREUDENTHAL, H. Über die Klassen von Sphärenabbildungen. I. *Comp. Math.* 5 (1937), 299–314. [6]

HILTON, P. J. Calculating the homotopy groups of A_n^2-polyhedra. I. *Quart. J. Math.* (Oxford), (2), 1 (1950), 299–309. [8]

HILTON, P. J. Calculating the homotopy groups of A_n^2-polyhedra. II. *Quart. J. Math.* (Oxford), (2), 2 (1951), 228–40. [8]

HILTON, P. J. Suspension theorems and the generalized Hopf invariant. *Proc. Lond. Math. Soc.* (3), 1 (1951), 462–93. [6, 8]

HOPF, H. Abbildungsklassen n-dimensionaler Mannigfaltigkeiten. *Math. Ann.* 96 (1927), 209–24. [3]

HOPF, H. Topologie der Abbildungen von Mannigfaltigkeiten. 1. Neue Darstellung der Theorie des Abbildungsgrades. *Math. Ann.* 100 (1928), 579–608. [3]

HOPF, H. Topologie der Abbildungen von Mannigfaltigkeiten. 2. Klasseninvarianten. *Math. Ann.* 102 (1930), 562–623. [3]

HOPF, H. Über die Abbildungen der 3-Sphäre auf die Kugelfläche. *Math. Ann.* 104 (1931), 637–65. [6]

HOPF, H. Über die Abbildungen von Sphären auf Sphären niedrigerer Dimension. *Fundam. Math.* 25 (1935), 427–40. [5, 6]

HU, S. T. An exposition of the relative homotopy theory. *Duke J. Math.* 14 (1947), 991–1033. [2, 3, 4]

HU, S. T. Inverse homomorphisms of the homotopy sequence. *K. Ned. Akad. van Wet.* (Indagationes Math.), 9, 2 (1947), 3–11. [4]

HUREWICZ, W. Beiträge zur Topologie der Deformationen. I–IV. *Proc. K. Akad. van Wet. Amst.* 38 (1935), 112–19, 521–8; 39 (1936), 117–26, 215–24. [2, 3]

HUREWICZ, W. and STEENROD, N. E. Homotopy relations in fibre spaces. *Proc. Nat. Acad. Sci., Wash.*, 27 (1941), 60–4. [5]

MACLANE, S. and WHITEHEAD, J. H. C. On the 3-type of a complex. *Proc. Nat. Acad. Sci., Wash.*, 36 (1950), 41–8. [7]

PONTRJAGIN, L. S. Homotopy classification of mappings of the $(n+2)$-sphere into the n-sphere. *Doklady Akad. Nauk SSSR*, 70 (1950), 957–9. [6]

POSTNIKOV, M. M. 1. Determination of the homology groups of a space by means of homotopy invariants. 2. On the homotopy type of polyhedra. 3. On the classification of continuous mappings. *Doklady Akad. Nauk SSSR*, 1951, Tom 76, 3, 359–62, Tom 76, 6, 789–91, Tom 79, 4, 573–6. [7]

Serre, J.-P. Homologie Singulière des Espaces Fibrés. *Ann. Math.* 54 (1951), 425–505. [5, 7, 8]

Serre, J.-P. Sur la Suspension de Freudenthal. *C.R. Acad. Sci., Paris*, 234 (1952), 1340–2. [6, 8]

Steenrod, N. E. and Whitehead, J. H. C. Vector fields on the n-sphere. *Proc. Nat. Acad. Sci., Wash.*, 37 (1951), 58–63. [6]

Whitehead, G. W. A generalization of the Hopf invariant. *Ann. Math.* 51 (1950), 192–238. [2, 6]

Whitehead, G. W. On the $(n+2)$nd homotopy group of the n-sphere. *Ann. Math.* 52 (1950), 245–7. [6]

Whitehead, J. H. C. On adding relations to homotopy groups. *Ann. Math.* 42 (1941), 409–28. [2, 6]

Whitehead, J. H. C. On $\pi_r(V_{n,m})$ and sphere-bundles. *Proc. Lond. Math. Soc.* 48 (1944), 243–91. [4, 5, 6, 8]

Whitehead, J. H. C. The homotopy type of a special kind of polyhedron. *Ann. Soc. Polon. Math.* 21 (1948), 176–86. [8]

Whitehead, J. H. C. On simply-connected 4-dimensional polyhedra. *Comm. Math. Helv.* 22 (1949), 48–92. [7, 8]

Whitehead, J. H. C. Combinatorial homotopy. I. *Bull. Amer. Math. Soc.* 55, 3 (1949), 213–45. [7]

Whitehead, J. H. C. Combinatorial homotopy. II. *Bull. Amer. Math. Soc.* 55, 5 (1949), 453–96. [4, 7]

Whitehead, J. H. C. A certain exact sequence. *Ann. Math.* 52 (1950), 51–110. [7, 8]

INDEX AND GLOSSARY

(Definitions of terms are not always given here. For other definitions, see the relevant part of the text.)

Randall Library – UNCW
QA611 .H65 NXWW
Hilton / An introduction to homotopy theory.
304900208110S